Integrated Assessment of Health and Sustainability of Agroecosystems

Advances in Agroecology

Series Editor: Clive A. Edwards

Agroecosystems in a Changing Climate
Paul C.D. Newton, R. Andrew Carran, Grant R. Edwards, and Pascal A. Niklaus

Agroecosystem Sustainability: Developing Practical Strategies
Stephen R. Gliessman

Agroforestry in Sustainable Agricultural Systems
Louise E. Buck, James P. Lassoie, and Erick C.M. Fernandes

Biodiversity in Agroecosystems
Wanda Williams Collins and Calvin O. Qualset

Interactions Between Agroecosystems and Rural Communities
Cornelia Flora

Landscape Ecology in Agroecosystems Management
Lech Ryszkowski

Multi-Scale Integrated Analysis of Agroecosystems
Mario Giampietro

Soil Ecology in Sustainable Agricultural Systems
Lijbert Brussaard and Ronald Ferrera-Cerrato

Soil Organic Matter in Sustainable Agriculture
Fred Magdoff and Ray R. Weil

Soil Tillage in Agroecosystems
Adel El Titi

Structure and Function in Agroecosystem Design and Management
Masae Shiyomi and Hiroshi Koizumi

Tropical Agroecosystems
John H. Vandermeer

Advisory Board

Integrated Assessment of Health and Sustainability of Agroecosystems

Thomas Gitau
Margaret W. Gitau
David Waltner-Toews

CRC Press
Taylor & Francis Group
Boca Raton London New York

CRC Press is an imprint of the
Taylor & Francis Group, an **informa** business

CRC Press
Taylor & Francis Group
6000 Broken Sound Parkway NW, Suite 300
Boca Raton, FL 33487-2742

International Standard Book Number-13: 978-1-4200-7277-8 (Hardcover)

Library of Congress Cataloging-in-Publication Data

Library of Congress Cataloging-in-Publication Data
Gitau, Thomas.
Integrated assessment of health and sustainability of agroecosystems / Thomas Gitau, Margaret W. Gitau, David Waltner-Toews.
p. cm. -- (Advances in agroecology)
Includes bibliographical references and index.
ISBN 978-1-4200-7277-8 (hardback : alk. paper)
1. Agricultural ecology. 2. Sustainable agriculture. 3. Agroforestry. 4. Agrobiodiversity. 5. Soil ecology. 6. Tillage. I. Gitau, Margaret W. II. Waltner-Toews, David, 1948- III. Title. IV. Series.

S589.7.G48 2008
630--dc22 2008013094

Visit the Taylor & Francis Web site at
http://www.taylorandfrancis.com

and the CRC Press Web site at
http://www.crcpress.com

To all those who made Thomas Gitau's work possible, and to those who supported him and his family as he took his final journey.

Foreword

This welcome volume on agricultural sustainability is the twelfth in the series addressing 'Advances in Agroecology' that has been published by CRC Press since 1997. It is authored by two innovative scientists both well-known in their fields, Dr. Margaret W. Gitau of Florida A&M University and Professor David Waltner-Toews of the University of Guelph, Canada. The research described in the book is based on the work of Thomas Gitau, who led a study in the ecological sustainability and impacts on human health of tropical highland agroecosystems in Kenya. His premature death at the age of 38 in 2005 was a great loss to the scientific community and particularly to the International Development Research Centre that funded his work. Fortunately, his work was almost complete and the publication of this volume was made possible by the efforts of Dr. Margaret Gitau, Dr. Thomas Gitau's widow, who was very much involved in his research, and Professor Waltner-Toews, who was a partner in the research.

The book is unique in providing links between human health and well-being issues and agroecosystem ecological sustainability, and draws on both management theory and practice, as well as participatory action research and ecologically-based complex systems theories. The work addresses problems that are crucial to many rapidly changing and urbanizing tropical environments. It represents an in-depth case study of a tropical agroecosystem that produced results and that has global ramifications in relation to sustainability.

Many of the tools used by Dr. Gitau were developed from other fields and incorporated in more innovative ways than work reported in many comparative books. It is clearly written and provides a great deal of practical information as well as addressing theoretical and systems issues in depth.

Clive A. Edwards
Professor of Entomology and Environmental Science
Soil Ecology Laboratory
The Ohio State University

Preface

About the same time as the work described by Thomas Gitau in this volume was being designed, the Ecosystem Approaches to Human Health Programme Initiative (also known as the Ecohealth Programme) of Canada's International Development Research Centre (IDRC) was exploring health and environment research as an eco- and socially systemic set of reflective actions to produce sustainable changes in the health of human populations. Working with colleagues at the International Livestock Research Institute (Nairobi) and the Universities of Guelph and Nairobi, Gitau proposed an exciting mix of qualitative participatory action and systemic thinking to address agroecosystem health issues in the Kenyan Highlands. The work intrigued IDRC with its innovative use of qualitative participatory approaches in a formal, systemic, assessment framework drawn from environmental and health risk assessments. IDRC was also very interested in the integration of development activities into research processes since the prevailing approach to development at the time was to separate intervention from research. Such an approach often seemed to lead to lost opportunities for learning on the one hand and to ill-conceived interventions on the other.

At the time, applying the concept of health to ecosystems was controversial. The value-laden nature of health was felt by some to detract from objective observation and modeling of ecosystem phenomena and to be unscientific. There were concerns that ecosystem health was anthropomorphic or at least anthropocentric, and that it would distract researchers and policymakers alike from the urgent necessity of ecological conservation. Yet, the concept of ecosystem health held traction with decision makers and appealed to a broader audience than the ideas of resilience and integrity favored by many ecologists. Some forward-thinking ecological thinkers, such as Allen and Hoekstra (1991), began to argue that people and their processes were part of ecosystems, and that the problems of ecological sustainability were of a scale and complexity that required not only multidisciplinary scientific approaches but also increasing participation of civil society.

Meanwhile, theoreticians such as James Kay, David Waltner-Toews, and David Rapport were building on the work of Henry Regier and others to develop a systemic framework for applying ecosystem health. Community and decision maker participation, interventions, and monitoring were integrated into the research process. To these researchers, the health metaphor facilitated the integration of human dimensions into ecological thinking, an element they felt was crucial to making meaningful progress toward changing how people relate to environment to improve ecological sustainability.

Thomas Gitau's research was among the earliest to apply ecosystem health concepts in an African rural development context. His work was instrumental in demonstrating the feasibility of fully participatory action-oriented research meant to understand the dynamics of an agricultural ecosystem. He and his collaborators formalized the collaborative and participatory characterization of the agroecosystem and the identification of indicators with scholarly applications of systems theory, social and political sciences, and ecology. He had a unique way of adapting methods from other disciplines, such as ecology, when no techniques were apparent within the discipline at hand for answering important questions. He was both rigorous and open-minded.

At IDRC, community-based natural resource management thinking was evolving to consider how more sustainable natural resources management could lead to better human health and well-being. IDRC Ecohealth Programme Initiative staff found the health paradigm to be easily understood and conceptually flexible in initial diagnoses of health problems with environmental drivers. Thomas Gitau's systems descriptions became known as "ecosystem mapping," a tool now widely advocated by IDRC's Ecohealth Programme officers in the development of new projects. The participation of civil society, government, and business stakeholders in the diagnosis of environment and health problems and in the design and implementation of interventions based on strong research findings is a central tenet of the Ecohealth Programme. Many subsequent applications of approaches similar to Thomas Gitau's have shown that actions to establish a healthier community must be community led to achieve meaningful and lasting impact.

Challenges identified by Thomas are still relevant today. There are methodological difficulties to transdisciplinary approaches. There are political impediments to the empowerment of disadvantaged and sometimes marginalized groups. Technical methods for holistic data analysis are still imperfect. Many current IDRC projects still grapple with some or all of these, and yet progress is still made.

Today, there is the learned International Society for Ecology and Health and an academic journal called *EcoHealth*. There are communities of practice contributing both theoretically and empirically to the evolution of the approach. Perhaps most telling are the increasing number of concrete examples of positive changes in the world as a result of ecosystem health thinking; ecological and social improvements in several downtown Kathmandu wards where animals had been slaughtered (a project that had begun as an attack on a simple disease problem [hydatid disease]) grew into a decade-long series of resilient local organizational, environmental, and health changes (e.g., Waltner-Toews et al., 2005). Many of the techniques used had been developed by Gitau in Kenya. Similarly, work on linking child health and natural resource use in the Peruvian Amazon drew strongly on Gitau's work (Murray et al., 2002; Murray, 2000). In Mexico's manganese mining districts, ecohealth approaches have successfully been applied to define a health problem, raise community and local government awareness, and pursue the translation of research findings into policy changes to protect human health (Montes et al., 2008; Riojas-Rodriguez et al., 2006). In Malawi, researchers working closely with the community helped introduce soyabean to the cropping system, with ensuing substantial impacts on child nutrition. The project used ecohealth approaches to develop locally appropriate uses for the beans while exploring a variety of social factors influencing child nutrition (Bezner Kerr et al., 2007). IDRC today continues to develop and apply systemic and integrated action research on environment and health problems using transdisciplinary multistakeholder approaches with attention to gender and social equity.

Much of the theoretical and practical work that Gitau's thinking influenced and interacted with has also been summarized and synthesized in two books. *Ecosystem Sustainability and Health: A Practical Approach* (David Waltner-Toews, 2004,

Cambridge University Press) is built around the development of the Adaptive Methodology for Ecosystem Sustainability and Health, to which Gitau's work was a major contributor. *The Ecosystem Approach: Complexity, Uncertainty, and Managing for Sustainability* (edited by David Waltner-Toews, James Kay, and Nina-Marie Lister, Columbia University Press, 2008) includes essays by colleagues working in Nepal, India, Peru, New Zealand, various sites in Africa, and North and South America, as well as a chapter by Gitau and his co-worker.

The reader of this book will learn much about the practical realities of implementing participatory, multistakeholder, transdisciplinary ecosystem health research. The book presents tools and approaches that can be applied in many rural development research projects today. And, the reader will certainly benefit from Thomas Gitau's qualities as a researcher and a convener. Thomas Gitau was a pioneer whose scientifically rigorous and conceptually innovative research in Kiambu contributed to the early development of ecosystem health thinking. Having briefly known Thomas while we were both graduate students, I can also say that he was a kind and generous soul whose untimely death robbed us not only of his scientific potential but also of the pleasure of his congenial company.

REFERENCES

Allen, T.F.H., and Hoekstra, T.W. (1991). *Toward a Unified Ecology.* Columbia University Press, New York. 384 pp.

Bezner Kerr, R., Dakishoni, L., Shumba, L., Msachi, R., and Chirwa, M. (2007). "We Grandmothers Know Plenty: Breastfeeding, complementary feeding, and the multifaceted role of grandmothers in Malawi." *Social Science and Medicine* December 20 [Epub ahead of print].

Montes, S., Riojas-Rodríguez, H., Sabido-Pedraza, E., and Ríos, C. (2008). Biomarkers of manganese exposure in a population living close to a mine and mineral processing plant in Mexico. *Environmental Research* 106(1): 89–95.

Murray, T. (2000, April). *Investigating Health, Biodiversity, and Natural Resource Use on the Amazon Frontier.* IDRC Bulletin. Available at: http://www.idrc.ca/es/ev-5276-201-1-DO_TOPIC.html.

Murray, T.P., Kay, J.J., Waltner-Toews, D., and Raez-Luna, E.F. (2002). Linking human and ecosystem health on the amazon frontier. In *Conservation Medicine: Ecological Health in Practice.* Aguirre, A., Ostfeld, R.S., Tabor, G.M., House, C., and Pearl, M.C., eds. New York: Oxford University Press, chapter 23.

Riojas-Rodríguez, H., Ríos, C., Rosas, I., Sabido Pedraza, E., Miranda, J., Siebe, C., Texcalac, J.L., and Santos-Burgoa, C. (2006). Motor alterations associated with exposure to manganese in the environment in Mexico. *Science of the Total Environment* 368(2–3): 542–556.

Waltner-Toews, D., Neudoerffer, C., Joshi, D.D., and Tamang, M.S. (2005). Agro-urban ecosystem health assessment in Kathmandu, Nepal: Epidemiology, systems, narratives. *EcoHealth Journal* 2(2): 155–164.

Dominique Charron
Program Leader, Ecosystem Approaches to Human Health Program
International Development Research Centre, Ottawa, Canada
January 2008

Acknowledgments

Without the enthusiastic support and participation of communities in Kiambu district, this work would not have been possible. We are eternally indebted to them. Deep gratitude is expressed to Dr. John McDermott for being a friend and a teacher. His encouragement and support nourished the academic endeavors in his work. In addition, Professors Joseph Gathuma and Erustus Kang'ethe were instrumental in the design and implementation of the project. Their friendship, advice, and encouragement were vital, more so during some difficult times in the course of this work.

We are grateful to the International Development Research Centre (IDRC) for providing most of the funds that supported this work. Special thanks to Dr. Bertha Mo and Don Peden, the program officers responsible for this grant. Their support and facilitation are greatly appreciated. Thanks also to the University of Nairobi for providing funds and material support to conduct this study. The Department of Public Health, Pharmacology, and Toxicology provided facilities and a conducive atmosphere for research. Many of the academic staff were generous with valuable advice.

The District Commissioner of Kiambu provided us with enthusiastic support, as did the District Development Committee. The assistance of officers—too many to mention by name—from the Ministries of Agriculture, Health, and Public Works in Kiambu district is appreciated with thanks. The district extension staff in Kiambu district deserve special mention.

Thanks to Musembi Musyoki and Margaret Kitonga-Lubaale for their advice on participatory action research techniques; Inga Kimaru, Eva Ochola, Rosemond Cheboi, Ruth Mwikali, James Mwangi, and Rose Kariuki for their diligence as research assistants; and Professor Kang'ethe for facilitating a number of village workshops and for sharing his insight on participatory action research techniques with us. Mr. K.A. Marimba handled all the logistics for the project—a job splendidly done. Special thanks to Dorcas Chege for typing, computing, and data support to the project.

Thanks also to all members of the multidisciplinary team, including Virginia Kimani, Julius Kilungo, Amos Omore, Maina Mbogo, Lucy Mwaura, Reuben Muni, Collette Suda, Sanyisha Khasiani, Kamau Mubuu, Jeane Mbugua, and Githaiga Kamau. Dr. John Curry and Mr. Jon Turner provided some invaluable advice to the multidisciplinary team.

The Authors

Thomas Gitau was born in Nairobi, Kenya, on February 28, 1967. An epidemiologist by profession, Thomas worked in the Department of Public Health, Pharmacology, and Toxicology at the University of Nairobi, where he taught graduate and undergraduate courses in analytical epidemiology and statistics. Thomas also served as a consultant for the International Livestock Research Institute (ILRI) from 2002. He held a bachelor degree of veterinary medicine (1992), a master of science degree in veterinary clinical studies (1994), and a doctoral degree in epidemiology and biostatistics, all from the University of Nairobi. During the course of his doctoral work, Thomas also received nondegree training from the Ontario Veterinary College at the University of Guelph in Canada. Thomas was a pioneer in the development and application of the agroecosystem health approach. His research work has been extended to other areas, including Uganda, Zambia, and beyond. Sadly, Thomas passed away on March 8, 2005, after a long battle with leukemia bravely borne.

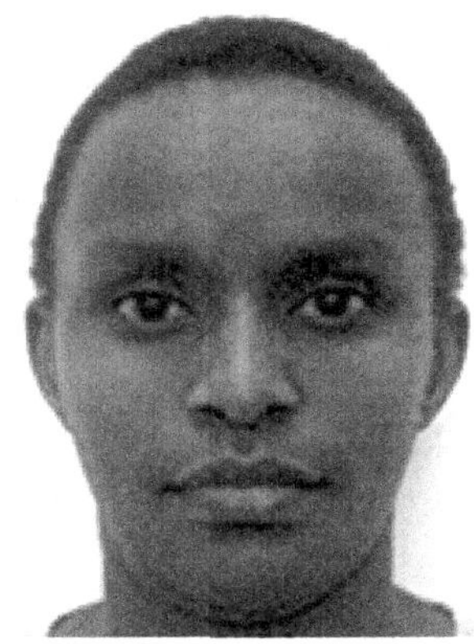

Margaret W. Gitau specializes in environmental and natural resources engineering and management. She has extensive experience and expertise in nonpoint source pollution, including fate and transport of agricultural pollutants, ecological modeling, and integrated approaches to developing workable and cost-effective management solutions. Dr. Gitau's research interests include Geographic Information Systems applications, development of computational technologies, land use trends and impacts analyses, and socio-economic aspects of environmental and natural resource conservation and management. Dr. Gitau received her B.S. and M.S. degrees in agricultural engineering from the University of Nairobi, and a Ph.D. in agricultural and biological engineering from the Pennsylvania State University. She is an assistant professor at Florida A&M University where she teaches courses in nonpoint source pollution, natural resources conservation engineering, and water management. Dr. Gitau serves as an associate editor for the *American Society of Agricultural and Biological Engineers*, and as a reviewer for several internationally recognized journals.

David Waltner-Toews is a veterinary epidemiologist, essayist, poet, and fiction writer. A professor of epidemiology at the University of Guelph, Ontario, Canada, he specializes in diseases people acquire from animals and in ecosystem health. He is the founding president of Veterinarians Without Borders/Vétérinaires sans Frontières—Canada (www.vwb-vsf.ca), as well as of the Network for Ecosystem Sustainability and Health (www.nesh.ca). Beside being an author of about 100 peer-reviewed scholarly papers, he has published half a dozen books of poetry, an award-winning collection of short stories, a murder mystery, and four books of nonfiction, ranging from the serious and scholarly to popular science, with titles such as *Food, Sex and Salmonella*, and *The Chickens Fight Back*. His most recent texts include *Ecosystem Sustainability and Health: A Practical Approach* (Cambridge University Press, 2004) and *The Ecosystem Approach: Complexity, Uncertainty, and Managing for Sustainability*, coedited with James Kay and Nina-Marie Lister (Columbia University Press, 2008).

Overview

A process was designed to assess the health and sustainability of a tropical highland, smallholder-dominated agroecosystem. Twelve study sites within the agroecosystem were selected in a multistage, purposive sampling protocol. Six of the study sites were designated *intensive* (ISSs). In these, some agroecosystem health and sustainability remedial measures were instituted. The other six study sites were designated *extensive* (ESSs) and were used to cross-validate the indicators and to provide statistical power.

Communities in the ISSs were included in the health and sustainability assessment. Participatory action research methods were used for that purpose. Human activity systems were modeled and analyzed using soft systems methods. Relationships among agroecosystem and sustainability factors were explored using loop models, graph theory, and pulse process models. Conventional observational study methods were used to study land-use units (LUUs).

Two sets of health and sustainability indicators were developed. One set—the community-driven suite of indicators—was developed by the communities in the ISSs as a list of measures that would help them assess their agroecosystem. These indicators were used to develop a community-based agroecosystem health and sustainability monitoring system. The other set of indicators—the research-based suite—was developed by a multidisciplinary team of researchers. Multiple correspondence analysis was used to further refine this suite of indicators and to develop a basis for their interpretation.

Although the process used in this study was similar in some ways to traditional approaches in research and development, there was an important departure in that communities were part of the analytical processes. In addition, the entire process was grounded in a unifying theoretical background that facilitated holistic analysis. More important, however, is that communities were able to use the concept of health to discuss and model approaches to improve their livelihoods. The approach provided a simple, yet highly specialized language—understood by the communities, researchers, extension agents, development agents, and policymakers—for discussing issues of health and sustainability of agroecosystems and for structuring the process through which remedial actions could be undertaken.

Research-based indicators differed in several important aspects from the community-driven ones. Researcher-proposed indicators focused mostly on numeric, non-value-based measures. This suite had a dearth of suitable social, and less so economic, indicators. In contrast, community-based indicators were more strongly value based, focusing mostly on a social-economic interpretation of the underlying biophysical phenomena. The community-based suite contained many indictors suitable for assessing many of the attributes in the social and economic domains. The suites were found to be complementary, with researchers requiring some of the data gathered using the community indicators and vice versa. Because of the short span of the project, it is difficult to assess the construct validity of the indicator suites.

However, the fact that communities, policymakers, and researchers were using information derived from these indicators in making decisions about Kiambu suggests that these measures are useful. In addition, several remedial actions taken as a result of monitoring using these indicators seem to be a move toward sustainability and better agroecosystem health.

Multiple correspondence analysis was found to be a useful method of summarizing and presenting data from indicators. Graphical techniques in conjunction with simple conceptual models were also found to be useful. A meaningful assessment of health and sustainability of Kiambu would require longitudinal studies over several years. However, time-dynamic models can be used to project trends in some of the key agroecosystem health attributes under various possible scenarios. The use of pulse process models for this purpose was explored and found potentially useful.

Goals and objectives of farmers and communities indicate a strong affinity to farming. With an average per capita farm income of 1,339.77 ± 179.43 shillings per annum (U.S. \$17.63 ± 2.36), this affinity did not seem to be based on the economic returns but more on sociocultural affinity to farming. In contrast, the average monthly wage was 6,537.11 ± 1,179.47 shillings (U.S. \$86.02 ± \$15.51), although only an average of 16% of the adults in a household were wage earners. The key constraint to health appeared to be mainly related to infrastructural and policy inadequacies. With effort, communities were able to make changes, and some of the problems were solved. At this stage, however, these appeared to be more of a reaction to immediate needs rather than long-term strategies.

Community inertia was mostly attributed to an inability of communities to influence decision making and policy. Development agenda was seen as being based on considerations other than the needs and aspirations of the communities. In addition, there were both sociocultural and legal impediments to communities setting—and working toward—reasonable goals for their own agroecosystems. Examples are the regulation of coffee and tea production and marketing, centralized planning and management of health services, and lack of community involvement in the maintenance of the road networks. Another reason could be that, in most cases, community expectations far outstripped the outcomes for a given objective. An example is Githima village, where construction of additional classrooms was expected to result in increased literacy levels during the same time span as it takes increased farm productivity to result in increased incomes. Such imbalances can only lead to a great deal of frustration and inertia. On the other hand, the ease with which communities were able to construct detailed cognitive maps and take some remedial actions suggests the existence of a collective understanding and capacity for consensus building and collective action. Coupled with the fact that communities were highly receptive to the concepts of action research, collective planning, monitoring, and evaluation, this could be interpreted as indicative of a great potential for improved health and sustainability for these communities given certain institutional and policy changes as well as expert support.

Contents

Foreword ... vii
Preface ... ix
Acknowledgments ... xiii
The Authors ... xv
Overview ... xvii

Chapter 1 General Introduction ... 1

1.1 Introduction ... 1
1.2 Global Context ... 2
1.3 Sustainability ... 4
1.3.1 Definitions ... 4
1.3.2 Assessment and Implementation ... 5
1.4 Agroecosystem Health ... 5
1.4.1 Systems Theory ... 7
1.4.2 The Health Concept ... 8
1.4.3 Assessment and Implementation ... 10
1.4.4 Indicators ... 11
1.4.5 Selection of Indicators ... 12
1.5 The Kiambu Agroecosystem ... 13
1.5.1 External Environment ... 13
1.5.2 Internal Environment ... 14
1.6 Potential Indicators ... 15
1.7 Justification ... 16
References ... 17

Chapter 2 Design and Implementation of an Adaptive, Integrated Approach to Health and Sustainability in a Smallholder-Dominated Agroecosystem ... 25

2.1 Introduction ... 25
2.2 Research Strategy and Methods ... 27
2.2.1 Secondary Data and Holarchical Scales ... 29
2.2.2 Sampling Study Sites ... 30
2.2.3 Systemic Description and Action Planning ... 31
2.2.4 Indicators ... 32
2.2.4.1 Community-Driven Indicators ... 32
2.2.4.2 Selection of Research-Based Indicators ... 32
2.2.5 Monitoring, Evaluation, and Assessment ... 33
2.2.5.1 Community-Based Assessments ... 33
2.2.5.2 Research-Based Assessments ... 33

2.2.6 Implementation of Interventions ... 33
2.2.6.1 Community Training ... 34
2.2.6.2 Community-Based Development Interventions ... 34
2.3 Results ... 34
2.3.1 Holarchical Scales ... 36
2.3.2 Study Sites ... 37
2.3.3 Systemic Description ... 38
2.3.3.1 Demographic Features ... 38
2.3.3.2 Geoclimatic Features ... 38
2.3.3.3 Resource Use and Distribution ... 38
2.3.3.4 Agriculture ... 42
2.3.3.5 Agroecosystem Health Goals ... 48
2.3.4 Health and Sustainability Assessment ... 49
2.3.5 Implementation of Interventions ... 49
2.4 Discussion ... 49
2.4.1 Holarchical Scale ... 49
2.4.2 Systemic Description ... 51
2.4.3 The Agroecosystem Health Approach ... 52
2.4.4 Health and Sustainability Assessment ... 53
References ... 54

Chapter 3 Community Participation and the Integration of Agroecosystem Health and Sustainability Concerns into Practical Decision Making ... 57

3.1 Introduction ... 57
3.2 Process and Methods ... 60
3.2.1 Community Identities ... 60
3.2.2 Initial Contact with Communities in the Study Sites ... 61
3.2.3 Initial Village Participatory Workshops ... 61
3.2.4.1 Participatory Techniques ... 63
3.2.4.2 Participatory Mapping ... 63
3.2.4.3 Institutional Mapping ... 64
3.2.4.4 Historical Background ... 65
3.2.4.5 Time Lines ... 65
3.2.4.6 Trend Lines ... 65
3.2.4.7 Transect Walks ... 66
3.2.4.8 Semistructured Interviews ... 66
3.2.4.9 Seasonal Calendars ... 66
3.2.4.10 Daily Activity Charts ... 67
3.2.4.11 Focus Group Discussions ... 67
3.2.4.12 Presentations and Analysis ... 67
3.2.4.13 Wealth Ranking ... 68
3.2.4.14 Health Analysis ... 68
3.2.4.15 Problem Identification and Ranking ... 68
3.2.4.16 Problem Analysis ... 69

3.2.4.17 Preference Ranking 69
3.2.4.18 Action Planning 69
3.2.5 Follow-up 69
3.2.5.1 Creating Cognitive Maps 70
3.2.5.2 Developing Community-Based Indicators 70
3.2.5.3 Monitoring, Evaluation, Planning, and Assessments 70
3.3 Results 70
3.3.1 Community Identities 71
3.3.1.1 Participatory Mapping 71
3.3.1.2 Institutional Mapping 71
3.3.1.3 Historical Background and Time Lines 71
3.3.2 Profiles and Trends 72
3.3.2.1 Trend Lines 72
3.3.2.2 Transect Walks and Semistructured Interviews 76
3.3.2.3 Wealth and Well-Being 76
3.3.2.4 Health Analysis 80
3.3.3 Problem Analysis and Action Planning 81
3.3.3.1 Problem Identification, Ranking, and Analyses 81
3.3.3.2 Preference Ranking and Action Planning 82
3.3.4 Follow-Up 82
3.3.4.1 Collective Action 82
3.3.4.2 Reflection and Replanning 82
3.4 Discussion 84
3.4.1 Community Identities and Collective Action 87
3.4.2 Community Participation 87
3.4.3 Systemic Descriptions 88
3.4.4 Problem Analyses and Action Plans 89
3.4.5 Monitoring and Evaluation 89
References 89

Chapter 4 Use of Systems Theory, Directed Graphs, and Pulse Process Models in an Adaptive Approach to Agroecosystem Health and Sustainability 91

4.1 Introduction 91
4.2 Process and Methods 93
4.2.1 Participatory Cognitive Mapping 94
4.2.2 Geometric Analyses 95
4.2.3 Pulse Process Models 96
4.2.4 Application of System Theory Tools in Villages 98
4.3 Results 98
4.3.1 Githima 99
4.3.2 Gitangu 104
4.3.3 Kiawamagira 109
4.3.4 Mahindi 110

4.3.5 Gikabu ... 113
4.3.6 Thiririka ... 117
4.4 Discussion ... 122
4.4.1 Construction of Cognitive Maps ... 122
4.4.2 Use of Signed Digraphs ... 122
4.4.3 Geometric Analyses ... 123
4.4.4 Pulse Process Models ... 124
4.4.5 Assessment of Value Stability ... 124
References ... 125

Chapter 5 Soft System Methodology in the Management of Agroecosystem Health and Sustainability Concerns of a Tropical Highlands Agroecosystem ... 127

5.1 Introduction ... 127
5.2 Process and Methods ... 129
5.2.1 Examination of the Problem Situations ... 129
5.2.2 Root Definitions and Soft System Modeling ... 129
5.2.3 Building Consensus, Compromise, and Collective Action ... 130
5.3 Results ... 130
5.3.1 Problem Situations ... 130
5.3.1.1 Drainage and Access Road Problems in Kiawamagira ... 130
5.3.1.2 Gitangu Water Project ... 131
5.3.1.3 Inadequate Extension Services in Kiambu District ... 134
5.3.1.4 Community Inertia in Kiambu District ... 136
5.3.2 Building Consensus and Root Definitions ... 136
5.3.3 Implementation, Monitoring, and Evaluation ... 137
5.4 Discussion ... 139
5.4.1 Soft System Models ... 139
5.4.2 Collective Action, Action Research, and Soft System Methodology ... 143
References ... 145

Chapter 6 Development of Health and Sustainability Indicators for a Tropical Highlands Agroecosystem ... 147

6.1 Introduction ... 147
6.2 Process and Methods ... 148
6.2.1 Development of Community-Driven Indicators ... 149
6.2.2 Development of Researcher-Proposed Indicators ... 151
6.2.3 Indicator Measurements ... 152
6.2.3.1 Community-Driven Indicators ... 152
6.2.3.2 Researcher-Proposed Indicators ... 152
6.2.4 Refining Researcher-Proposed Indicators ... 153
6.3 Results ... 153
6.3.1 Community-Driven Indicators ... 153
6.3.2 Researcher-Proposed Indicators ... 155

6.3.3 Indicator Measurement and Refinement 158
6.3.3.1 Community Driven 158
6.3.3.2 Researcher Proposed 159
6.3.4 Comparison of Indicator Suites 167
6.4 Discussion 170
6.4.1 Comparison of Indicator Suites 170
6.4.2 Indicator Measurement and Refinement 174
6.4.3 Practicality and Application 177
References 178

Chapter 7 An Assessment of Health and Sustainability of a Smallholder-Dominated Tropical Highlands Ecosystem 181

7.1 Introduction 181
7.2 Process and Methods 182
7.2.1 Spatial and Temporal Trends in the Indicators 182
7.2.2 Evaluation of Goals, Expectations, and Achievements 183
7.3 Results 184
7.3.1 Spatial Trends in the Researcher-Proposed Indicators 184
7.3.2 Evaluation of Temporal Trends in the Researcher-Proposed Indicators 185
7.3.3 Evaluation of Goals, Expectations, and Achievements 187
7.4 Discussion 190
7.4.1 Spatial and Temporal Trends in the Indicators 190
7.4.2 Evaluation of Goals, Expectations, and Achievements 193
References 196

Chapter 8 General Discussion 197

8.1 Synopsis 197
8.2 Sustainability 197
8.3 Agroecosystem Health 198
8.4 Kiambu Agroecosystem 199
8.5 Health and Sustainability Assessment 200
8.6 Summary 201

Chapter 9 General Conclusions 203

Appendices 205
Appendix 1: Questionnaire Used to Carry Out a Census of All the Land-Use Units in the Village 206
Appendix 2: Questionnaire Used for Indicator Measurement at the Land-Use Unit Level 208

Index 223

1 General Introduction

1.1 INTRODUCTION

Kenya's food security depends on the ability to increase agricultural productivity (Yudelman, 1987) without degrading further—but rather facilitating the regeneration of—the resources on which agriculture depends. How can increases in productivity be achieved and sustained? Many technologies have been demonstrated to increase agricultural productivity. What is becoming increasingly clear is that many of these may not be sustainable, mostly because they engender degradation of the resources on which agriculture and human well-being depend. The question is therefore not so much how to increase agricultural productivity but how optimal productivity can be achieved and sustained.

The central highlands agroecosystem in Kenya serves as a good example of how conventional technology-based approaches to agricultural productivity can result in failure, reemergence of old problems, and development of new ones. Efforts have been geared toward maximizing off-take per unit area (Delgado, 1989) through intensification of land use (Winrock International, 1992) and increased use of external input and technologies. The result has been a proliferation of intensively farmed smallholder units—now the dominant land-use system in the highlands. This transformation has had limited success as well as important failures. In some cases, there were initial increases in productivity, but many are now registering declines, attributed mostly to land degradation and disintegration of the traditional balance among people, their habitat, and economic systems (Mohamed-Saleem and Fitzhugh, 1995). The realization that smallholder agriculture depends on a complex of interrelated sociocultural and biophysical factors has led to their description as complex, diverse, and risk prone (Chambers et al., 1989).

While causes of technology failure are not always obvious, it is clear that conventional methods are severely limited in their ability to deal with the complexity of systems such as smallholder farming in the East African highlands. Sustainable transformation of such systems requires an adaptive and integrated approach—one that takes a systems perspective, incorporates holistic views of well-being, and takes into account the multiple goals and multiple perspectives of the primary managers of these systems. Issues of human values (such as economics and aesthetics), scale, and discipline (environmental, economic, social, etc.) are central and must be accentuated and solved rather than obscured (Waltner-Toews, 1996). In addition, technical feasibility and economic viability must not be the only criteria for evaluating new strategies (Woomer, 1992). Other criteria, such as social and environmental costs, efficacy, efficiency, and effectiveness, must also be included. It is this view that has led to the articulation of a new outlook in agricultural development, embodied in concepts such as sustainability and agroecosystem health.

Sustainable agriculture has been defined as the successful management of resources for agriculture to satisfy changing human needs while maintaining or enhancing the quality of the environment and conserving natural resources (Technical Advisory Committee, 1987). An agricultural system that is sustainable must be resource conserving, socially supportive, commercially competitive, and environmentally sound (Ikerd, 1990). It allows the demands for food and other products to be met at a socially acceptable economic and environmental cost (Crosson, 1993). In a sustainable system, agricultural activities would have little or no adverse effects on their ecosystem and yet remain gainful (in terms of profits and other utilities) to the producers themselves and to the wider social organization to which they belong (Lynam, 1993).

In spite of an expanding ecological and economic literature on sustainability, the concept has remained largely inoperative in applied research (Izac and Swift, 1994). The main obstacle has been that the current definitions of sustainable agriculture, although attractively holistic, are too vague and ambiguous to lead to clear-cut measurements of the sustainability of specific agroecosystems (Izac and Swift, 1994).

It has been suggested that uncertainties inherent in holistic assessments can be reduced by relying on trends in a group of carefully chosen attributes (Rapport and Regier, 1980; Rapport, 1992). Measures of such attributes or their proxies—known as *indicators*—assessed over time and space can provide an objective assessment of sustainability. The agroecosystem health approach provides a framework through which indicators of sustainability can be selected and measured.

Given this background, the general objective of this study was to carry out an integrated assessment of agroecosystem health and sustainability with special focus on smallholder farms in Kiambu District, Kenya. Specifically, the study aimed at

1. Adapting the agroecosystem health framework for use in a smallholder-dominated tropical highlands agroecosystem.
2. Developing a suite of health and sustainability indicators for smallholder farms in the Kiambu agroecosystem.
3. Using the selected indicators to assess health and sustainability of the systems.
4. Enabling farmers and communities to assess the health and sustainability of their own agroecosystems.
5. Assessing the potential of various strategies in improving the health and sustainability of the agroecosystem.

1.2 GLOBAL CONTEXT

The world's population more than doubled over the decades 1950 to 1990, increasing from 2.5 billion in 1950 to 5.3 billion in 1990 (Lynam, 1993). Because of this, most of the earth's resources have had to be commanded for agricultural production. Agriculture has become the most expansive land-use system in the world. Consequently, it is a major determinant of the quality and quantity of other natural resources, such as fresh water, forests, grasslands, and undomesticated plant and animal life (Lynam, 1993).

As most of the world's resources became engaged, the capacity for expansion of agriculture diminished greatly. Attention shifted to intensification—with the aim of maximizing productivity per unit of limiting resource. The result was yield-maximizing technologies based mainly on fossil energy and fossil-derived chemicals. The increase in productivity has been so tremendous that Europe, North America, and several other parts of the world are now confronted with the problem of surplus production (Treitz and Narain, 1988). Owing to its high yield potential, this "high-input agriculture" has been rapidly and widely adopted, replacing many traditional agricultural practices.

Although the world now produces more food per capita than at any other time in history (Waltner-Toews, 1996), agriculture has failed to satisfy the needs of a big section of the world's population. The reasons for these are twofold. The first is that agricultural productivity is highly heterogeneous, following patterns that widely differ from those of population density (Pimental and Hall, 1984). These differences are becoming more pronounced since resource-poor regions also tend to have the highest population growth rate. Second, today's agriculture requires heavy subsidies (Pimental and Hall, 1984), implying that resource-poor regions cannot achieve expected yields from technologies based on it.

Sub-Saharan Africa (SSA) is one region in the world where food demand is far higher than the current production capacity (Lal, 1987; Okigbo, 1990; Brown and Thomas, 1990). This has been attributed to many factors, including resource scarcity, high population growth rate, and the inability to adopt fully high-input technology. Three other important constraints are social disruption due to wars and urbanization, climatic changes (Okigbo, 1990), and severe environmental degradation.

In SSA, decline in agricultural productivity and land degradation appear to be locked in a vicious cycle driven by the spiraling population pressure. The high population growth rate ensures that demand for agricultural produce remains high, while more resources are required for purposes other than agriculture. Traditional practices (e.g., crop rotation, leaving land fallow), which for centuries have been used to ensure natural resource preservation, have been disrupted (Okigbo, 1990; Yudelman, 1987). Increases in agricultural production have been achieved through increasing the percentage of land under cultivation to include marginal areas, forest reserves, and hill slopes. Millions of hectares of land have been cleared for food production, in most cases without consideration of the ecological consequences. The result of this process is frightening: land degradation, erosion, silting rivers (Treitz and Narain, 1988), poverty, hunger, and malnutrition.

These and similar outcomes in other parts of the world point to what are now growing causes of concern: (1) Most of the resources on which today's agriculture depends are nonrenewable, and (2) agricultural practices are major contributors to environmental degradation. That many technological innovations—although having high yield-maximizing potential—have only served to exacerbate environmental degradation and carry unacceptable social costs is widely accepted. The most serious global concern during the twenty-first century will be to feed people without destroying the natural resource base (Treitz and Narain, 1988). Focus has to turn to agricultural practices with a potential for maintaining optimum productivity over time (P.A. Allen and Van Dusen, 1988). Successful management of agricultural and ecological systems will be seen as the ability to conserve or even enhance the resource base while meeting the reasonable needs of the people dependent on them.

1.3 SUSTAINABILITY

The term *sustainable development* was coined in the early 1980s and is now the major subject in development research all over the world (World Commission on Environment and Development, 1987). *Sustainable agriculture* has been used to embody the idea and goals of sustainability in agricultural production.

1.3.1 Definitions

Agricultural sustainability has been defined and characterized in vastly different ways (Harrington, 1992). Each of the many definitions is devised from a different perspective and for a different purpose (Winograd, 1994), and little headway has been made in the search for a comprehensive and concise definition (Pearce et al., 1990). The various definitions can be categorized into one or more of three main perspectives: the agroecological concept, the resource concept, and the growth concept (Harrington, 1992).

Definitions with an agroecological perspective focus on sustainability in terms of *system resilience*. This is the ability of an agricultural system or ecology to maintain its productivity when subject to stress or perturbation (Conway, 1986). In this sense, sustainability of a system is enhanced through system diversity and increased efficiency in use and recycling of nutrients and energy (Altieri, 1987). Consequently, monitoring trends in system diversity and in the internal cycling of nutrients and energy would be fundamental in an assessment of sustainability (Harrington, 1992).

The second category of definitions places emphasis on *stewardship* or the proper care and protection of resources (Barker and Chapman, 1988). According to this perspective, the sustainability of agriculture can best be enhanced by slowing economic development, stabilizing human population levels, and discouraging the exploitation of natural resources (Barbier and McCracken, 1988; Durning, 1990). According to this view, measuring sustainability involves an assessment of the quantity and quality of natural resources available now and in the future.

In the last category are definitions that focus on the need for continued growth in agricultural productivity while maintaining the quality and quantity of the resources devoted to agriculture (Technical Advisory Committee, 1987). This requires that renewable resources be used at a rate lower than that at which they can be regenerated, wastes be emitted at a rate lower than that at which they are absorbed by the environment, and that use of nonrenewable resources be optimized (Barbier and McCracken, 1988).

Despite the many differing definitions, there are some notions common to all (Harrington, 1992). One such notion is that measuring sustainability implies drawing conclusions or stating probabilities about future events. All such forecasts contain varying degrees of uncertainty. The degree to which sustainability can be measured depends on the degree of accuracy of predictions about the future. Another idea common to all concepts of sustainability is that a measure of sustainability is based on a time frame. However, different time frames apply to different components of sustainability (Harrington, 1992). Some problems, such as soil nutrient depletion, are best studied over the medium term. Some, such as erosion and salinization, are best studied over longer time frames (Harrington, 1992).

Last, sustainability can be realized (and measured) at several different levels. In agriculture, such levels could include the plot, the farm, village, catchment, geopolitical or geoclimatic zones, national, and finally global. The various levels are nested within each other, forming a conceptual hierarchy of concentric layers. Although the sustainability of a specific level in the hierarchy is directly related to the functional state of the sublevels within it, not all the sublevels need be sustainable. Some resources may be used in excess of sustainable levels in some units, and the overall sustainability of the system may be maintained by substituting among resources and between sublevels over time (Graham-Tomasi, 1991).

1.3.2 Assessment and Implementation

The inability to find a concise definition of sustainability has been viewed as the obstacle in integrating sustainability concerns in practical decision making (Pearce et al., 1990; Izac and Swift, 1994; Lynam and Herdt, 1989; Graham-Tomasi, 1991). It has been argued that scientific research necessitates refining holistic concepts such as sustainability to a more specific and rigorous one (Izác and Swift, 1994; Pearce et al., 1990). The countering argument is that various components and factors in such complexes as agricultural systems have extensive, complex, and dynamic interrelationships. Any activity or reaction therefore has a high degree of unpredictability on both the spatial and temporal scales (Holling, 1986, 1992). Furthermore, the action-reaction chain flows—in most cases—in the stochastic, nondeterministic, and often counterintuitive fashion characteristic of soft systems. Attempts at assessing agricultural sustainability as if it is an objectively verifiable state of a nonhierarchical hard system can therefore only result in a great deal of frustration and confusion (Waltner-Toews, 1996).

The use of indicators to assess sustainability is a widely accepted approach (Swindale, 1992; Izac and Swift, 1994; Winograd, 1994; van Bruschem, 1997; Aldy et al., 1998; Smit et al., 1998). There have been attempts to develop generic indicators of sustainability. Such processes have been complicated by the fact that sustainability issues are system and scale specific. The choice of indicators and their interpretation depend largely on the context in which they are used. What is needed to implement the broad ideas about sustainability is not so much another list of indicators to measure, but an integrated framework within which such indicators can be developed and interpreted (Waltner-Toews, 1991).

1.4 AGROECOSYSTEM HEALTH

Ecosystem health is an emerging science paralleling human and veterinary medicine with, as its goal, the systematic diagnosis and treatment of stressed agroecosystems (Schaeffer, 1991). It extends and modifies the concept of sustainable agriculture (Waltner-Toews, 1994) to provide a systematic method for diagnosis, prognosis, and rehabilitation of agricultural ecosystems (Rapport, 1995).

The term *ecosystem*, coined in the mid-1930s, was first defined as the collection of all organisms and environments in a single location (Tansley, 1935). With the understanding of the interrelationships between the biotic and the abiotic elements in a geographic location, the term came to denote an organizational unit that includes one or more living entities through which there is a transfer and processing of energy and matter (Evans, 1956). More recently, an ecosystem has been defined as a functional system of complementary relations between living organisms and their environment, delimited by arbitrarily chosen boundaries, which in space and time appear to maintain a steady yet dynamic equilibrium (Gliessman, 1990). An agricultural ecosystem (*agroecosystem*) is a similar conceptual construct used to describe parts of the biosphere managed primarily for the purpose of agricultural production (Waltner-Toews, 1996).

The largest ecosystem is the biosphere—the portion of the earth that is populated by living things. Within it are many smaller ecosystems of varying dimensions and with different structures and functions, forming a hierarchical complex of systems (Bossel and Bruenig, 1989). The biosphere is to a great extent a closed system with regard to matter but an open system with regard to energy. The ecosystems within it, however, rarely act as closed systems; there is a continuous circulation of materials and energy within and between them.

A characteristic feature of natural ecosystems is their ability—within limits—to regulate themselves (Tivy and O'Hare, 1981). This results from the extensive interactions among the various abiotic and biotic components of an ecosystem to form complex feedback loops. The feedback signals responsible for such homeostatic control are material phenomena (Clapham, 1983), such as flows of energy, nutrients, and metabolic wastes.

Human beings are the ecologically dominant species in nearly all ecosystems of the world (Tivy and O'Hare, 1981). Through various technological advancements, they manipulate ecosystems to favor their survival above that of all other species. At the highest level of manipulation, natural ecosystems are completely disarticulated and totally replaced by new assemblages of plant and animal species (Toledo, 1990). Agricultural ecosystems (agroecosystems) are an example of human-manipulated ecosystems, with the aim of favoring agricultural production.

In an agroecosystem, only a few species are allowed to exist, while other species are removed through expenditures of energy, herbicides, and pesticides. The abiotic environment is controlled through extensive use of irrigation, fertilization, and tilling. Agroecosystems would change their forms if farmers were not able to generate and maintain a series of signals that counterbalance the natural successional forces (Clapham, 1983). In comparison with natural ecosystems, agroecosystems are therefore intrinsically unstable, requiring continuous inputs of energy, materials, and technology from human sources to maintain their integrity (Toledo, 1990). They tend to be relatively simple, comprising a suite of populations that would not normally constitute a natural ecosystem (Clapham, 1983). This reduction in biological diversity implies a simplification of trophic structure, while many niches are left unoccupied (Gliessman, 1990).

As the interactions between the various ecosystem components are disrupted, the flow of many feedback signals either ceases or is greatly modified, suspending many of the homeostatic mechanisms (Rappaport, 1971; Pimental and Hall, 1984). In addition, new avenues for nutrient and energy losses—such as export of produce, leaching, and erosion—develop (Gliessman and Amador, 1980). Socioeconomic considerations become the most significant determinants of the functional state of the resulting ecosystem.

Despite the control of ecosystem processes that the farmers maintain, agroecosystems are in a constant state of flux as they respond to such other influences as climatic changes and geographic variation (Clapham, 1983). Managing agroecosystems effectively requires an understanding of the signals generated within an ecosystem and how these signals—in their totality—influence the various functions and states of the ecosystem (Clapham, 1983).

1.4.1 Systems Theory

A system has been defined as an abstract concept of a whole (Checkland et al., 1990), consisting of a group of parts or components that interact according to some kind of process (Odum, 1983) and behave as a whole in response to stimuli applied to any of the parts (Spedding, 1988). To distinguish this from the common usage of the word system (real-world arrangement of things or processes), Checkland (1988) and Checkland and Scholes (1990) suggested the use of an alternate term: *holon*. This term was coined by Koestler (1978), who spoke of reality as Janus faced, like the two-faced Roman god. He referred to each unit (person, organism, etc.) as a holon and the nested hierarchy of which they are a part as a *holarchy*. The interactive combination of parts within a holon confers new properties to the system over and above those of the individual components that constitute it. Properties arising from interactions between parts—termed *emergent properties* (Checkland et al., 1990)—are only apparent when taking an overview of the system.

An ecosystem can be described as a holon that exhibits the emergent property of having a capacity to regulate and organize its own internal structure and function and to mitigate stresses imposed from outside (Rapport et al., 1985). This property—termed *integrity* (Kay, 1991)—imparts to the holon a capacity to perpetuate itself over time even within a fluctuating environment. Sustainability would therefore result when a holon has the capacity to maintain both its integrity and its productivity over all the foreseeable fluctuations.

A distinctive feature of ecosystems is that they can be described as occurring in nested hierarchies (Waltner-Toews and Wall, 1997), where entities at different scales are nested within each other in concentric layers. The nested hierarchies form holarchies, with each nested entity considered a holon (Checkland, 1981). Several holarchies can be described for agroecosystems, depending on the features (e.g., ecological, cultural, social, or economic) used to delineate the holons within it. For example, a biophysical holarchy can be defined as consisting of fields nested within farms, catchments, watersheds, drainage basins, agroecozones, and larger bioregions. A socioeconomic holarchy can be conceptualized as individuals nested within households, villages, larger administrative or sociopolitical boundaries, all the way to the global community. Each level in a holarchy has its own emergent properties. It contributes to the nature of, and is affected by, levels above and below it. Each level of the agroecosystem is therefore a subsystem of a bigger ecosystem that is in turn part of a wider environment (Kay, 1994).

A holon that has integrity must possess both monitoring and control structural or functional relations between its components. Monitoring is an assessment of the performance of the system. The monitoring subunit integrates signals that indicate changes in either the internal or the external conditions. The control subunit provides the mechanisms through which the holon can adapt to the new conditions (Checkland et al., 1990). Monitoring and control therefore require that there be at least one measure of performance and a definition of what constitutes good or bad performance (Checkland et al., 1990).

Performance can be judged based on three general criteria: efficacy, efficiency, and effectiveness. *Efficacy* is when the processes involved are adequate to produce the required output in sufficient quantities and with the required quality. *Efficiency* is when the minimum possible resources are utilized during the process. If the various outputs resulting from the activities within the holon are consistent with the purposes and state of the larger whole in which the holon is a part of, then the criteria of *effectiveness* is satisfied.

1.4.2 The Health Concept

For centuries, scholars and practitioners in the health sciences have struggled with questions of physical abilities and disabilities, self-perceptions, remedies, and their assessment (Waltner-Toews, 1996). Similar questions are asked of agroecosystems today. Furthermore, the general methodologies for screening, diagnosis, risk assessment, and fitness determination have been tested and applied in the health sciences for decades (Waltner-Toews, 1996).

The health metaphor proceeds from the view that health is an objectively definable state that, once described, can be effectively pursued. Health in the agroecosystem context depends on more than biophysical integrity; it is equally dependent on healthy socioeconomic processes, healthy human communities, and, more importantly, adaptive public policy (Rapport, 1995). Whether an agroecosystem is healthy is therefore a socioeconomic judgment as well as a biophysical assessment. This implies a degree of consensus among the stakeholders regarding what is a satisfactory definition of ecosystem health (Bergeron et al., 1994).

In ecosystems, as in organisms, what constitutes health depends not only on objective scientific criteria but also on subjective evaluation and value judgment (Rapport et al., 1985) and is partly reflective of sociocultural phenomena (Labonte, 1991). The definition of health therefore varies and will continue to vary for different units of concern, over time, and among cultures (Kark, 1979; Costanza et al., 1992). Likewise, lists of health attributes will vary with different conceptions of health. However, there may be attributes common to all, such as the notions of a harmonious balance and the notion of capacity to achieve a purpose (Last, 1987).

Productivity, stability, equitability, and self-reliance are some of the health attributes that have been described (Conway and McCracken, 1990; Gallopin, 1994a, 1994b). Health attributes are seen as emergent properties of agroecosystems. The relative degree to which an ecosystem shows any of these attributes contributes to the overall perception of its health. Indicators would be those measurable parameters that would be expected to change with changes in these attributes.

Productivity (Conway and McCracken, 1990) refers to the range, value, quality, and quantity of products derived from the agroecosystem. Izac and Swift (1994) distinguished three types of products that communities derive from an agroecosystem. The first is harvestable yield, which includes crops, livestock products, medicine, and the like that farmers deliberately grow. The second type includes amenities—environmental services provided by the agroecosystem such as drinking water, fuel, and an aesthetically pleasing environment. The third type includes by-products, those material outcomes, beneficial or detrimental to the farming communities' well-being, that are the consequence of the process of production and amenity use such as soil erosion and water pollution.

Stability refers to the agroecosystem's response to perturbation. The term stability encompasses several different properties of the ecosystem (Rutledge, 1974), many of which have been variously described (Holling, 1973; Orians, 1975; Cairns and Dickson, 1977; Robinson and Valentine, 1979; Harrison, 1979; van Voris et al., 1980). The most comprehensive is the description by Orians (1975), which identifies seven properties related to stability: constancy, persistence, inertia, elasticity, amplitude, cyclical stability, and trajectory stability. *Constancy* is lack of change in some parameters of the system. *Persistence* is its survival time, while *inertia* is the ability to resist external perturbations. *Elasticity* refers to the rate at which the system returns to its former state following a perturbation. The magnitude of normal system oscillations is its *amplitude*. *Cyclical stability* is the property of a system to cycle about some central point or zone, while *trajectory stability* is the property of a system to move toward some end point or zone despite differences in the starting points.

Equitability is defined as the evenness of distribution of agroecosystem resources and amenities among the stakeholders (Conway and McCracken, 1990). Its importance is based on the value judgment that an egalitarian distribution is preferable, and that poverty is likely to force some stakeholders to use unsustainable practices (Izac and Swift, 1994). A feature of natural environments is that when products and amenities are supplied to one group of individuals, they are also available to other groups that were not the intended target (Pearce et al., 1990). This inability to exclude some stakeholders extends to the by-products of various agroecosystem processes, such as air and water pollution. Furthermore, exhaustion of natural resources such as soil nutrient depletion and erosion has major implications on intergenerational equity. At the farm household level, gender equity is an important social factor.

Self-sufficiency is where local and regional subsistence is derived mainly from the agroecosystem. Self-sufficiency is related to diversification in the sense that production systems must be diversified to satisfy the subsistence needs of all the stakeholders. More ecogeographical units are utilized, providing more opportunities for integration and combination of various production processes and the recycling of nutrients, energy, and wastes. Subordination of surplus production in favor of subsistence encourages resource conservation and a multiuse strategy.

The goal in a multiuse strategy is to obtain a maximum number of necessary products that each ecogeographical component offers and to maintain this over time. This favors two desirable (in a value judgment sense) ecological characteristics:

spatial heterogeneity and biological diversity (Toledo, 1990). Self-sufficiency is only crucial to sustainability in more or less closed systems. Open systems that have many avenues for the flow of energy and materials among them can collectively achieve sustainability through trade-offs, maximizing the use of renewable resources and minimizing the use of nonrenewable ones within each system. Those products that are too costly (economically, environmentally, or socially) to produce within one unit are obtained from another, in a process of mutual exchange.

Other health attributes include vigor, resilience, integrity, and adaptability. The *vigor* of a system is simply a measure of its activity (Costanza et al., 1998). *Resilience* refers to the system's ability to maintain its structure and behavior in the presence of stress (Holling, 1986). Cairns and Dickson (1977) identified four properties of ecosystems that determine their stress recovery characteristics: vulnerability, elasticity, inertia, and resiliency. They defined *vulnerability* as inability to resist irreversible damage. *Elasticity* is the ability to recover after displacement of structure or to a steady state closely approximating the original, while *inertia* is the ability to resist such displacement. *Resiliency* is the number of times a system can undergo the same disturbance and still snap back. If a system is able to maintain its organization in the face of changing environmental conditions, then it is said to have *integrity* (Kay, 1991). Organization of a system refers to the number and diversity of interactions between its components (Costanza et al., 1998). Integrity is therefore a composite property, tying together other characteristics such as stability, resilience, and vigor. *Adaptability* has been defined as the ability to undergo adaptive changes in response to change in the environment (Ho and Saunders, 1979).

1.4.3 Assessment and Implementation

Agroecosystem health assessment and implementation are carried out in five iterative steps: (1) describing the system of interest; (2) identifying the owners, actors, and customers; (3) setting or naming the goals and objectives of the system; (4) identifying and implementing feasible and desirable changes; and (5) monitoring appropriate indicators and reassessing the situation (Bellamy et al., 1996; Waltner-Toews and Nielsen, 1997). The agroecosystem health approach is complicated by three main conceptual dilemmas. First, agroecosystems, like all complex phenomena, can be viewed from a variety of perspectives, yet none of these can be labeled as right or wrong or as good or bad. For example, a systemic description from an economic perspective would not necessarily be analogous, comparable, or equivalent in any way to that reflecting an ecological perspective of the same agroecosystem (Waltner-Toews et al., 2000).

The second dilemma emanates from the fact that agroecosystems are holarchical systems (Allen and Hoekstra, 1992). Each level is a holon; that is, it is simultaneously a whole with its own emergent properties, comprised of smaller wholes, while itself a part of a bigger holon. Conceivably, health and sustainability at a level, say n, of a holarchy depends on trade-offs and balances among the holons in its penultimate layer ($n - 1$), implying that some holons may need to be unhealthy or unsustainable within specific spatiotemporal bounds to maintain the health and sustainability of the overall system. The third dilemma is that agroecosystems seek to optimize multiple goals, and yet—because of the human activity component in them—the goals are often competing and sometimes conflicting. Thus, the goal-seeking and self-organizing behaviors of agroecosystems occur in a series of trade-offs and balances with inherent contradictions.

An agroecosystem health assessment is undertaken to help people make better decisions regarding managing the agroecosystems in which they live and grow food (Waltner-Toews et al., 2000). It follows that the perspectives of the primary managers of agroecosystems are the most managerially useful descriptions of the agroecosystem. Furthermore, these descriptions incorporate in them the value judgments, goals, and objectives of the primary managers of the system. Recent developments in participatory (Chambers, 1989, 1994) and action research (Greenwood and Levin, 1998; Stringer, 1999) methods provide means through which farmers and other members of the community in an agroecosystem can be involved in the process.

Soft systems methodology, developed by management specialist Peter Checkland (Checkland and Scholes, 1990), provides a systems approach to the management of complex situations in which (1) multiple perspectives exist, (2) there is no consensus on what the problem is, (3) no single solution can be agreed on, and (4) multiple competing or conflicting goals exist. Soft systems methodology can be used—in combination with participatory and action research methods—to build a community-centered process that resolves the issues of multiple perspectives and multiple goals within an agroecosystem health research process.

The issue of scale and trade-offs among holons within and between levels is difficult to resolve. At what scale should an agroecosystem health assessment be carried out? Which levels and which units must be healthy and sustainable, and which must be traded off? Focusing on particular scales may lead to inappropriate conclusions regarding lower or higher levels in the hierarchy. The choice of scale should be guided by the questions that initiated the concern for health and sustainability. For example, concern about the health and sustainability of a smallholder-dominated agroecosystem calls for attention at the farm level and the community or watershed level. The latter level is because this is where there is a degree of integration in terms of social and economic factors, and the former is because most decisions are made at this level. Izac and Swift (1994) propose that to understand sustainability at one level, there is a need to understand the level above and the one below.

1.4.4 Indicators

In human and animal health, the diagnostic process involves taking measurements on specific parameters and comparing them with ranges in a healthy individual. Ecosystem health proposes a similar approach by which the spatiotemporal trends of health attributes or their proxies—known as *indicators*—are assessed (Rapport and Regier, 1980; Odum, 1983; Rapport et al., 1985; Swindale, 1992; Izac and Swift, 1994; Winograd, 1994; van Bruschem, 1997; Aldy et al., 1998; Smit et al., 1998). Gallopin (1994a) described an indicator as a variable and defined a variable as an operational representation of an attribute of a system. A variable has a set of possible outcomes, by which thresholds, standards, and targets are several such outcomes singled out because of their special relevance to the condition of the system.

Indicators must have a defined range in the healthy ecosystem (Schaeffer et al., 1988). In some cases, these ranges—described as thresholds, standards, and targets—depend on subjective value judgment (Gallopin, 1994a). The health status of an ecosystem is indicated not only by the comparison of indicators to such ranges but also by the pattern exhibited by the indicators over time and space and in relation to each other. Relationships between indicators are key to understanding ecosystem function and in relating cause to effect.

While the human and animal health diagnosis involves a limited set of parameters, ecosystems present a list that not only varies with level in the ecological hierarchy, but also consists of virtually an infinite number of measurable parameters (Schaeffer et al., 1988). Another difficulty has been that researchers from different disciplines, conditioned by particular perceptions of this complex situation and bound by habit to particular scales of research, arrive at different kinds of lists of parameters to be measured (Waltner-Toews, 1996). How can indicators be selected to ensure that the suite is at the same time parsimonious, covers all domains and all important attributes of the system, provides managerially useful information, and is cost-effective and timely to measure?

Since the aim of agroecosystem health research is the management of agroecosystems, it follows that the most useful suites of indicators are those that aid managers in their decision-making processes. Indicators must therefore be related to the goals and objectives of the agroecosystems as well as the capacity and potential of the system and the perceived risks and potential stresses. Although the goals and objectives may be based—to a large extent—on value judgment and can be highly subjective, measures of health (no matter what health is conceived to be) must be technical, multidisciplinary, and objective (Waltner-Toews and Wall, 1997). Ideally, indicators should be features of the agroecosystem that change with alteration in the health status of the ecosystem. Because of this, indicators have been referred to as the *vital signs* of the ecosystem (Rapport et al., 1985). Indicators may also be features of an ecosystem that indicate the presence, absence, or magnitude of stress or risk. These can be termed *risk factors*. A third category of indicators measures the potential, capabilities, or reserves of the ecosystem and can be termed *health promoters*.

1.4.5 Selection of Indicators

Lightfoot and Noble (1993) and Izac and Swift (1994) suggested different suites of indicators that can be used in the context of smallholder agroecosystems, while Thompson and Pretty (1996) used a number of indicators to assess the impact of a soil conservation program. Izac and Swift (1994) focused on the products, by-products, and amenities of the agroecosystem at various levels of the agricultural hierarchy. Although their list includes some elementary measures—such as soil pH, soil exchangeable aluminum content, and stream turbidity and acidity—that can be objectively assessed, most (such as nutritional status of households and communities, ratio of aggrading to degrading land area, and biodiversity and complexity) are compound attributes that can only be assessed using other proxy variables.

Thompson and Pretty (1996) included sustained increases in productivity, decreases in resource degradation, increases in local resilience, and decreases in vulnerability as indicators of soil conservation impact. They indicated that a participatory approach was used to elicit information but not how these data were objectively reorganized into such compound attributes as resource degradation, resilience, and vulnerability. Lightfoot and Noble (1993), based on a farming systems approach, focused on integration, efficiency, and recycling. Their list of indicators includes the number of interlinkages between and within systems, labor allocation, and the quantity of bioresources flowing between resource systems.

It can be argued that some "nonquantifiable" indicators provide more important information than more objective ones (Harrington, 1992). But, if the aim of a health assessment is to detect changes in the health status of agroecosystems, to compare one system with another, or to assess the potential impact of various factors on health, indicators must be amenable to an objective assessment. In addition, the choice of indicators must be tempered by practicality and the cost of measurement in terms of time and money. An assessment may be categorized as either descriptive or predictive of the system's health status (Ruitenbeek, 1991). The purpose of an assessment may be to assist management and decision making, set policy standards, determine policy compliance, or assess progress toward a goal (Boyle et al., 2001). The level of precision required may vary based on the purpose of the assessment as well as who the end user is.

1.5 THE KIAMBU AGROECOSYSTEM

1.5.1 External Environment

Only a third of Kenya is arable, and a shortage of suitable farmland is a severe constraint to the expansion of agricultural production. An ever-increasing human population further worsens this situation. In an attempt to increase per capita food production, two strategies have been sought (Food and Agricultural Organisation, 1981; Government of Kenya, 1983). One is to intensify production on cultivated lands; the other is to extend cultivation to marginal areas. It is now apparent that the low fertility of marginal areas allows little or no surplus to be produced (Mohamed-Saleem and Fitzhugh, 1995). In addition, these environments are too fragile to support more intensive and long-term agricultural production (Mwonga and Mochoge, 1989). Intensification in the zones of high agricultural potential remains the alternative with minimum environmental and social costs (Winrock International, 1992).

With a favorable high-altitude climate and a highly diversified agricultural system (Odingo, 1971), the Kenyan highlands are the most productive lands in the country. Scarcity of arable land, the high human population, and traditional practices continue to encourage subdivision of the farmlands into small units. Because of this and other sociocultural factors, smallholder farms dominate land use in the highlands and represent the largest farming population in the country (Woomer, 1992). In 1983, it was estimated that there were about 1.5 million smallholders in the country (Stotz, 1983).

Most of the land in smallholder farms is permanently under crops. This generally involves the continuous cultivation of maize (Ransom et al., 1995). Most smallholders in the highlands own and manage animals (Powell and Williams, 1993).

Ruminants (mainly cattle, but also sheep and goats) are the most important and are kept on the farm all year-round (Delgado, 1989). Hand hoeing, use of manure, intercropping, and sometimes mulching are common practices, but crop rotation, especially with perennial vegetation, is rare, and fallowing is not practiced. Only a few farmers retain small plots of grass for grazing purposes (Mati, 1989). Crop residues are used as fodder (Powell and Williams, 1995), especially for stall-fed cattle.

Traditionally, there was low population pressure, and labor was the major constraint—with its cost higher compared to land (Powell and Williams, 1995). Soil fertility was maintained through fallowing, which was preferred to manure because it required less labor (Stangel, 1995). The low inelastic demand for agricultural produce also ensured low demand for agricultural inputs. Productivity was increased—if required—by placing more land under cultivation. As the population increased, consumption patterns changed, and the land reserves became exhausted, the scope for these practices diminished greatly (Mohamed-Saleem and Fitzhugh, 1995). Despite this, technology substitutes have not been widely adopted. The result has been a decline in productivity in the long run due to loss of soil fertility (Ransom et al., 1995) and its basic chemical and physical coherence (Stangel, 1995). Some smallholder farmers now maintain constant or slightly increasing returns by increasing labor input (Mohamed-Saleem and Fitzhugh, 1995), but the continued use of traditional technologies to intensify land use results in declining returns.

The mixing of crop and livestock production in smallholder farms is important for several reasons. For one, mixed farming is the most viable agricultural enterprise on resource-poor and highly fragmented farm units (Delgado, 1989; McIntire et al., 1992; Winrock International, 1992) as it facilitates both recycling of nutrients and intensification of land use. The complement between crops and livestock produces a synergy between the two, thus increasing overall productivity (Davendra, 1993). The diversification spreads out the risk and increases the stability and resilience of the system while allowing higher labor input per unit of land (Delgado, 1989).

Agricultural intensification of smallholder agriculture, involving improved technology and inputs (McIntire et al., 1992), is inevitable (Mohamed-Saleem and Fitzhugh, 1995). In other developing regions, intensification has occurred gradually over many years, but in Africa it will need to happen over a very short time due to rapid population growth (Mohamed-Saleem and Fitzhugh, 1995). Some countries have relied on yield-increasing technologies based on fossil energy, but the costs of these are prohibitive to the majority of smallholder farmers in Kenya (Mohamed-Saleem and Fitzhugh, 1995). High-input technologies have also been associated with environmental pollution, while fossil energy is no longer considered a renewable resource. The challenge in Kenya is to increase per hectare and per animal yields by introducing yield-increasing and environmentally sound production innovations that are technically feasible and economically viable now and adaptable and sustainable into the future.

1.5.2 Internal Environment

Kiambu district comprises 2,500 km^2 of the central highlands. It is one of the most densely populated districts in the highlands, having an estimated density of 480 persons per square kilometer (Government of Kenya, 1994). Altitudes range from 1,400 m in the southeast to 2,400 m in the north. Rainfall is bimodal with the two peaks in April/May and October/November. Average rainfall is 1,100 mm per year. The most predominant soil type is nitisols (red Gikuyu loams). The combination of good soils, suitable climate, well-developed infrastructure, and the proximity to the country's main market (Nairobi) makes the district the most economic farming region in the country. Vegetable cultivation and dairy production (zero grazing) are the most feasible farming activities because of the small farm sizes and the high demand for produce in the city. Coffee, tea, pyrethrum, maize, beans, and bananas are also grown. Livestock are mainly dairy cattle of exotic breeds or their crosses with indigenous breeds.

The average farm size is 1.1 ha per household of 4.8 people (Jaetzold and Schmidt, 1983). Most of the land is devoted to crop production (Stotz, 1983), with the remainder used as a dwelling place for humans and housing for cattle. Very few smallholdings have pastures. Livestock are integrated with cropping activities under which crop by-products are used as fodder, while manure is used as fertilizer. The cattle are permanently housed (zero grazing) and hand-fed on fodder crops, crop residues, grass, and other material collected off the farm and commercial feeds. In most households, farming is a supplemental source of income since most households rely on off-farm employment for their main income.

There is a great potential to increase output per farm and per unit area of land in the mixed smallholder farms (Walshe et al., 1991). This can be achieved with the use of improved technology and inputs (McIntire et al., 1992). Hudson (1989) suggested that to increase yields, the primary requirement is not research into new methods, but the increased application of techniques and practices that are already known. Other requirements are improved crop varieties and livestock breeds, more research into local conditions, more use of fertilizers, more capital, more mechanization, and reduction of wastage from pest and disease.

1.6 POTENTIAL INDICATORS

From a national point of view, the goals of the Kiambu agroecosystem would include the stability and productivity of rural livelihoods, supplying wholesome produce for Nairobi, and producing cash crops (coffee and tea) for export. While these underlie the perceived efficacy and effectiveness of the Kiambu agroecosystem, other health and sustainability attributes depend on internal, systemic perspectives of its goals and objectives and how these relate to key biophysical and socioeconomic factors of the agroecosystem. Any list of indicators that does not consider this is not likely to be functional in the practical decision-making processes of the agroecosystem. What are these goals, and what are the key biophysical and socioeconomic factors? What attributes are more important in defining the health of the Kiambu agroecosystem? An agroecosystem assessment must first seek to address these questions before setting out lists of things to measure. It follows that indicators are part of the results of the process rather than the driver.

A distinguishing feature of Kenyan smallholder agriculture is the existence of the village as an important level in the agricultural hierarchy. At this level, there is a significant integration of ecological, economic, and social factors. Furthermore, the sociopolitical boundaries of the village often roughly correspond to an ecosystem at the catchment scale. Their social organization forms a unit in which farmers' practices are homogeneous, and economic activities are complementary. The village level provides information on external effects such as market mechanisms that operate above the farm level (Izac and Swift, 1994). This implies that a comprehensive list of indicators would be stratified into village-level and farm- or land-use unit-level indicators.

1.7 JUSTIFICATION

Is smallholder farming sustainable—and why? What are the effects of smallholder farming on the health of the highlands agroecosystem? What management strategies maintain or even enhance the health and sustainability of these agroecosystems? These questions demand an answer that goes beyond economic viability and technical feasibility. They demand systemic approaches that weigh all the costs: sociocultural, economic, and biophysical. They draw attention to the complex interrelationships of factors that govern agroecosystems, which in turn govern and sustain rural livelihoods. The significance of this is that "asking the right questions is the first step in finding useful answers."

Mixed crop-livestock smallholder farms are considered to have several advantages in terms of the health of the agroecosystem. The most important of these is their multiuse nature. Crops and livestock are integrated to create a system that is considered more efficient in the use of natural resources, requires less external input, and is more environmentally sound. Despite these advantages, yields in these farms are much lower compared to those obtained in experimental stations within similar agroclimatic zones. Researchers (e.g., Kilungo et al., 1994; Omore et al., 1994; Gitau et al., 1994; Kimani and McDermott, 1994) have identified several factors as the most important constraints to productivity, and it is widely accepted that there is potential for increased productivity. How can we identify the most suitable of all available strategies, and how can their implementation be managed and monitored?

Smallholdings are complex systems operating within a highly variable and diverse environment. They are mostly subsistent in nature, with low capital outlay and heavy dependence on environmental conditions. Many of them are managed as part of the traditional noncash economy, while the households that depend on them rely heavily on the growing cash-based system. This dichotomy means that the criteria for optimizing both the technical and allocative efficiency are often unclear even to the managers themselves. Furthermore, smallholders are unable to take advantage of the opportunities and information generated by the cash-based system, such as credit, and market data. The unwillingness or inability of smallholders to adopt and adapt new technologies further increases their dependency on the natural resource base. This means that many of them still rely on traditional methods of production, and yet the resource base and the production environment have changed to an extent that many of these methods are no longer viable. Much more critical, however, is the lack—among the smallholders—of critical information required to successfully husband the environment and resources on which their livelihoods depend. This has led to degenerative spirals in which poor resource management results in degradation and poverty, which in turn leads to the adoption of even more untenable resource management strategies.

Understanding the impact of different development strategies will require an adaptive, integrated, and systemic approach with both short-term and long-term monitoring and evaluation strategies. Farmers and communities—who are the primary managers of the agroecosystem—must be involved in the process both as local experts and as users of the local theory generated by the adaptive process. The agroecosystem health framework, incorporating the concept of sustainability and involving participatory, action research and soft systems methodologies, seems to be a suitable framework for the design and implementation of such a process.

Questions regarding the health and sustainability of smallholder-dominated agroecosystems go beyond surplus production and the viability of farm units. They are more fundamental: How should communities manage agroecosystems, not only to derive their livelihoods from them but also to conserve them and maintain them for posterity? These issues are broad and include human health and nutrition; employment; rural-urban migration; sociocultural capital; regional, gender, and intergenerational equity; and the environment. Tackling these requires a multifaceted approach incorporating policy, infrastructure development, and governance. An integrated, adaptive, and participatory process for assessing health and sustainability of an agroecosystem would therefore be a process to empower the people who live there, giving them command and control as the primary managers. Such an approach would provide them with the information and analytic capacity from which to negotiate goals and objectives, influence policy and demand services, as well as structure their collective actions toward better livelihood outcomes.

REFERENCES

Aldy, J.E., Hrubovcak, J., and Vasavada, U. (1998). The role of technology in sustaining agriculture and the environment. *Ecological Economics* 26: 81–86.

Allen, P.A., and Van Dusen, D. (1988). *Global Perspectives in Agroecology and Agriculture.* Agroecology Program, University of California, Santa Cruz. 730 pp.

Allen, T.H.F., and Hoekstra, T.W. (1992). *Toward a Unified Ecology.* Columbia University Press, New York. 384 pgs.

Altieri, M.A. (1987). *Agroecology: The Scientific Basis of Alternative Agriculture.* Westview Press, Boulder, CO. 227 pp.

Barbier, E., and McCracken, J. (1988). *Glossary of Selected Terms in Sustainable Economic Development.* IIED Gatekeeper Series SA7. International Institute for Environment and Development, London. 19 pp.

Barker, R., and Chapman, D. (1988). *The Economics of Sustainable Agricultural Systems in Developing Countries.* Working Paper 88-13. Cornell University, Ithaca, NY.

Bellamy, J.E.C., Waltner-Toews, D., and Nielsen, N.O. (1996). *Agroecosystem Health—A Fuzzy Systems Approach.* Discussion Paper 30. Agroecosystem Health Project, University of Guelph, Ontario, Canada. 20 pp.

Bergeron, G., Toledo, J., Tyrchniewicz, E., Weise, S., and Gameda, S. (1994). Health research: A comment. In *Agroecosystem Health with Special Reference to the Consultative Group for International Research (CGIAR): Proceedings of an International Workshop.* Nielsen, N.O., ed. Guelph, Ontario, Canada: Agroecosystem Health Project, University of Guelph, pp. 94–97.

Bossel, H., and Bruenig, E.F. (1989). Natural Resource Systems Analysis; Ecological and Socio-economic Systems and Sensitivity Analysis for the Conservation and Management of Forest Ecosystems and Natural Resources in South-East Asia. Report of a Training Course, 24 July–19 August 1989, Kuala Lumpur, Malaysia (English) Bossel, H., Bruenig, E.F./Deutsche Stiftung fuer Internationale Entwicklung, Feldafing (German). Zentralstelle fuer Ernaehrung und Landwirtschaft, 1989. 145 pp.

Boyle, M., Kay, J.J., and Pond, B. (2001). Monitoring in support of policy: an adaptive ecosystem approach. In *Encyclopaedia of Global Environmental Change* Volume 4, Munn, R., ed. New York: Wiley, pp. 116–137.

Brown, H.C.P., and Thomas, V.G. (1990). Ecological considerations for the future of food security in Africa. In *Sustainable Agricultural Systems*. Edwards, C.A., Lal, R., Madden, P., Miller, R.H., and House, G., eds. Ankeny, IA: Soil and Water Conservation Society, pp. 353–377.

Cairns, J.P., and Dickson, K. (1977). Recovery of streams from spills of hazardous materials. In *Recovery and Restoration of Damaged Ecosystems*. Cairns, J., Dickson, K.L., and Herricks, E.E., eds. Charlottesville: University Press of Virginia, pp. 24–42.

Chambers, R. (1989). Farmer first and TOT. In *Farmer First: Farmer Innovation and Agricultural Research*. Chambers, R., Pacey, A., and Thrupp, L.A., eds. London: Intermediate Technology, pp. 181–193.

Chambers, R. (1994). The origins and practice of participatory rural appraisal. *World Development* 22: 953–969.

Chambers, R., Pacey, A., and Thrupp, L.A. (1989). *Farmer First: Farmer Innovation and Agricultural Research*. Intermediate Technology, London. 219 pp.

Checkland, P. (1981). *Systems Thinking, Systems Practice*. Wiley, Chinchester, U.K. 344 pp.

Checkland, P., and Scholes, J. (1990). *Soft Systems Methodology in Action*. Wiley, Chinchester, U.K. 329 pp.

Checkland, P.B. (1988). The case for "holon." *Systems Practice* 1: 235–238.

Checkland, P.B., Forbes, P., and Martin, S. (1990). Techniques in soft systems practice part 3: Monitoring and control in conceptual models and in evaluation studies. *Journal of Applied Systems Analysis* 6: 29–37.

Clapham, W.B., Jr. (1983). *Natural Ecosystems*. Macmillan, New York. 282 pp.

Conway, G. (1986). *Agroecosystem Analysis for Research and Development*. Winrock International, Bangkok, Thailand. 111 pp.

Conway, G.R., and McCracken, J.A. (1990). Rapid rural appraisal and agroecosystem analysis. In *Agroecology and Small Farm Development*. Altieri, M.A., and Hecht, S.B., eds. Boca Raton, FL: CRC Press, pp. 221–235.

Costanza, R., Mageau, M., Norton, B., and Patten, B.C. (1998). Predictors of ecosystem health. In *Ecosystem Health*. Rapport, D., Constanza, R., Epstein, P.R., Gaudet, C., and Levins, R. eds. London: Batsford, pp. 240–250.

Costanza, R., Norton, B., and Haskell, B. (1992). *Ecosystem Health: New Goals for Environmental Management*. Island Press, Washington, DC. 269 pp.

Crosson, P. (1993). Sustainable agriculture: A global perspective. *Choices* 38–42.

Davendra, C. (1993). *Animal Production from Small Farm Systems in South-East Asia*. FAO Animal Production and Health Paper 106. Food and Agricultural Organisation, Rome.

Delgado, C.L. (1989). The changing economic context of mixed farming in savana West Africa: A conceptual framework applied to Burkina Faso. *Quarterly Journal of International Agriculture* 28: 3–4.

Durning, A. (1990). How much is enough?. *World Watch* 3: 12–19.

Evans, F.C. (1956). Ecosystem as the basic unit in ecology. *Science* 123: 1127–1128.

Food and Agricultural Organisation. (1981). *Agriculture: Towards 2000*. Food and Agricultural Organization, Rome.

Gallopin, G.C. (1994a). Agroecosystem health: A guiding concept for agricultural research? In *Agroecosystem Health with Special Reference to the Consultative Group for International Research (CGIAR): Proceedings of an International Workshop*. Nielsen, N.O., ed. Guelph, Ontario, Canada: Agroecosystem Health Project, Faculty of Environmental Sciences, University of Guelph, pp. 94–97.

Gallopin, G.C. (1994b). *Impoverishment and Sustainable Development: A Systems Approach*. International Institute for Sustainable Development (IISD), Winnipeg, Manitoba, Canada.

Gitau, G.K., O'Callaghan, C.J., McDermott, J.J., Omore, A.O., Odima, P.A., Mulei, C.M., and Kilungo, J.K. (1994). Description of smallholder dairy farms in Kiambu District, Kenya. *Preventive Veterinary Medicine* 21: 155–166.

Gliessman, S.R. (1990). Agroecology: Researching the ecological basis for sustainable agriculture. In *Agroecology: Researching the Ecological Basis for Sustainable Agriculture*. Gliessman, S.R., ed. New York: Springer-Verlag, pp. 3–10.

Gliessman, S.R., and Amador, M.A. (1980). Ecological aspects of production in traditional agroecosystems in the humid lowland tropics of Mexico. In *Tropical Ecology and Development*. Furtado, J.I., ed. Kuala Lumpur, Malaysia: ISTE, pp. 601–608.

Government of Kenya. (1983). *National Development Plan: 1984–1988*. Ministry of Planning and National Development, Nairobi.

Government of Kenya. (1994). *Kiambu District Development Plan: 1994–1996*. Rural Planning Department, Ministry of Planning and National Development, Nairobi.

Graham-Tomasi, T. (1991). Sustainability: Concepts and implications for agricultural research policy. In *Agricultural Resource Policy: International Quantitative Perspectives*. Pardey, P.G., Roseboom, J., and Anderson, J.R., eds. Cambridge, U.K.: Cambridge University Press, pp. 81–102.

Greenwood, D.J., and Levin, M. (1998). A history of action research. In *Introduction to Action Research: Social Research for Social Change*. Greenwood, D.J., and Levin, M., eds. Newbury Park, CA: Sage, pp. 15–31.

Harrington, L.W. (1992). Measuring sustainability: Issues and alternatives. In *Let Farmers Judge: Experiences in Assessing the Sustainability of Agriculture*. Hiemstra, W., Reijntjes, C., van der Werf, E., eds. London: Intermediate Technology, pp. 26–41.

Harrison, G.W. (1979). Stability under environmental stress: Resistance, resilience, persistence, and variability. *American Naturalist* 113: 659–669.

Ho, M.W., and Saunders, P.T. (1979). Beyond Neo-Darwinism—An epigenetic approach to evolution. *Journal of Theoretical Biology* 78: 573–591.

Holling, C.S. (1973). Resilience and stability of ecological systems. *Annual Review of Ecology and Systematics* 4: 1–24.

Holling, C.S. (1986). The resilience of terrestrial ecosystems, local surprise and global change. In *Sustainable Development of the Biosphere*. Clark, W.C., and Munn, R.E., eds. Cambridge, U.K.: Cambridge University Press, pp. 292–317.

Holling, C.S. (1992). Cross-scale morphology, geometry, and dynamics of ecosystems. *Ecological Monographs* 62: 447–502.

Hudson, N. (1989). *Soil and Water Conservation*. Batsford, London. 324 pp.

Ikerd, J.E. (1990). Sustainability's promise. *Journal of Soil and Water Conservation* 45: 4–14.

Izac, A.-M.N., and Swift, M.J. (1994). On agricultural sustainability and its measurement in small-scale farming in sub-Saharan Africa. *Ecological Economics* 11: 105–125.

Jaetzold, R., and Schmidt, H. (1983). *Farm Management Handbook of Kenya: Volume IIB—Natural Conditions and Farm Management Information, Central Kenya*. Ministry of Agriculture, Kenya, in cooperation with the German Agricultural Team (GAT) of the GTZ, Nairobi.

Kark, S. (1979). *Epidemiology and Community Medicine*. Random House, New York. 470 pgs.

Kay, J.J. (1991). A nonequilibrium thermodynamic framework for discussing ecosystem integrity. *Environmental Management* 15: 483–495.

Kay, J.J. (1994). *Some Notes on the Ecosystem Approach, Ecosystems as Complex Systems and State of the Environment Reporting*. University of Waterloo, Waterloo, Ontario, Canada.

Kilungo, J.K., McDemmott, J.J., Oluoch, W.K., and Omore, A.O. (1994). Socio-economics of smallholder dairy production in Kiambu District, Kenya. *The Kenya Veterinarian* 18: 247–249.

Kimani, V.W., and McDermott, J.J. (1994). Pesticide handling and usage by smallholder farmers on an irrigation scheme in the Central Province of Kenya. *The Kenya Veterinarian* 18: 540–542.

Koestler, A. (1978). *Janus: A Summing Up*. Random House, New York.

Labonte, R. (1991). Econology: Integrating health and sustainable development. Part One: Theory and background. *Health Promotion International* 6: 49–65.

Lal, R. (1987). Managing the soils of sub-Saharan Africa. *Science* 236: 1069–1076.

Last, J.M. (1987). *Public Health and Human Ecology*. Appleton and Lang, East Norwalk, CT. 464 pp.

Lightfoot, C., and Noble, R.P. 1993. A participatory experiment in sustainable agriculture. *Journal of Farming Systems Research Extension* 4(1): 11–34.

Lynam, J.K. (1994). Sustainable growth in agricultural production: The links between production, resources, and research. In *Opportunities, Use, and Transfer of Systems Research Methods in Agriculture to Developing Countries*. Goldsworthy, P., and De Vries, F.P, eds. Kluwer Academic Publishers Group, pp. 3–27.

Lynam, J.K., and Herdt, R.W. (1989). Sense and sustainability: Sustainability as an objective in international agricultural research. *Agricultural Economics* 3: 381–398.

Mati, B.M. (1989). Cultural and structural aspects of soil conservation: An evaluation report. In *Soil and Water Conservation in Kenya: Proceedings of the Third National Workshop*, Kabete, Nairobi, Kenya, 16–19 September 1986. Thomas, D.B., Biamah, and Kilewe, L., eds. University of Nairobi, Dept. of Agricultural Engineering, pp. 163–175.

McIntire, J., Bourzat, D., and Pingali, P. (1992). *Crop-Livestock Interactions in Sub-Saharan Africa*. World Bank, Washington, D.C. 246 pp.

Mohamed-Saleem, M.A., and Fitzhugh, H.A. (1995). An overview of demographic and environmental issues in sustainable agriculture in sub-Saharan Africa. In *Livestock and Sustainable Nutrient Cycling in Mixed Farming Systems of Sub-Saharan Africa*. Vol. 2. Powell, J.M., Fernández-Rivera, S., and Williams, T.O., eds. Addis Ababa: ILCA, pp. 3–20.

Mwonga, S., and Mochoge, B.O. (1989). The effects of cultivation on some physical and chemical properties of three Kenyan soils. In *Soil and Water Conservation in Kenya: Proceedings of the Third National Workshop*, Kabete, Nairobi, Kenya, 16–19 September 1986. University of Nairobi, Dept. of Agricultural Engineering.

Odingo, R.S. (1971). *The Kenya Highland: Land Use and Agricultural Development*. East African Publishing House, Nairobi.

Odum, E.P. (1985). Trends in Stressed Ecosystems. *Bioscience* 35: 419–422.

Odum, H.T. (1983). *Systems Ecology: An Introduction*. Wiley, Chichester, U.K. 644 pp.

Okigbo, B.N. (1990). Role of international organisations in developing sustainable production systems in sub-Saharan Africa: research priorities and programs. In *Proceedings of the Symposium on the Sustainability of Agricultural Production Systems in Sub-Saharan Africa*. International Conference on Livestock and Sustainable Nutrient Cycling in Mixed Farming Systems of Sub-Saharan Africa, Addis Ababa, Ethiopia, 22–26 November 1993. Vol. 2. Powell, J.M., Fernández-Rivera, S., and Williams, T.O., eds. Addis Ababa: ILCA.

Omore, A.O., McDermott, J.J., and Carles, A.B. (1994). Comparison of productivity of cattle grazing systems in small-holder dairy farms in Kiambu District, Kenya. In *Proceedings of the 7th International Symposium on Veterinary Epidemiology and Economics.* Nairobi, pp. 121–123.

Orians, G.H. (1975). Diversity, stability and maturity in natural ecosystems. In *Unifying Concepts in Ecology.* Van Dobben, W.H., and Lowe-McConnell, R.H., eds. The Hague, Netherlands: B.V. Publishers, pp. 139–150.

Pearce, D., Barbier, E., and Markandya, A. (1990). *Sustainable Development: Economics and Environment in the Third World.* Edward Elgar, London. 217 pp.

Pimental, D., and Hall, C.W. (1984). *Food and Energy Resources.* Academic Press, Orlando, FL. 268 pp.

Powell, J.M., and Williams, T.O. (1993). *Livestock, Nutrient Cycling, and Sustainable Agriculture in the West African Sahel.* Gatekeeper Series SA37. International Institute for Environment and Development, London. 15 pp.

Powell, J.M., and Williams, T.O. (1995). An overview of mixed farming systems in sub-Saharan Africa. In *Livestock and Sustainable Nutrient Cycling in Mixed Farming Systems of Sub-Saharan Africa.* Vol. 2. Powell, J.M., Fernández-Rivera, S., and Williams, T.O., eds. Addis Ababa: ILCA.

Ransom, J.K., Ojiem, J., and Anampiu, F.K. (1995). Nutrient flux between maize and livestock in a maize-coffee-livestock system in central Kenya. In *Livestock and Sustainable Nutrient Cycling in Mixed Farming Systems of Sub-Saharan Africa.* Vol. 2. Powell, J.M., Fernández-Rivera, S., and Williams, T.O., eds. Addis Ababa: ILCA, pp. 411–418.

Rappaport, R.A. (1971). The flow of energy in an agricultural society. *Scientific American* 225: 116–132.

Rapport, D.J. (1992). Evolution of indicators of ecosystem health. In *Ecological Indicators.* McKenzie, D.H., Hyatt, D.E., and McDonald, V.J., eds. London: Elsevier Applied Science, pp. 121–134.

Rapport, D.J. (1995). Ecosystem health: What's in a name? *Ecosystem Health* 1: 1–2.

Rapport, D.J., and Regier, H.A. (1980). An ecological approach to environmental information. *Ambio* 9, 22–27.

Rapport, D.J., Regier, H.A., and Hutchinson, T.C. (1985). Ecosystem behavior under stress. *American Naturalist* 125, 617–640.

Robinson, J.V., and Valentine, W.D. (1979). The concepts of elasticity, invulnerability and invadability. *Theoretical Biology* 81: 91–104.

Ruitenbeek, H.J. (1991). *Indicators of Ecologically Sustainble Development: Towards New Fundamentals.* Ministry of Supply and Services, Ottawa.

Rutledge, R.W. (1974). *Ecological Stability: A Systems Theory Viewpoint.* Oklahoma State University, Stillwater. 93 pp.

Schaeffer, D.J. (1991). A toxicological perspective on ecosystem characteristics to track sustainable development. *Ecotoxicology and Environmental Safety* 22: 225–239.

Schaeffer, D.J., Herricks, E.E., and Kerster, H.W. (1988). Ecosystem health: I. Measuring ecosystem health. *Environmental Management* 12, 445–455.

Smit, B., Waltner-Toews, D., Rapport, D., Wall, E., Wichert, G., Gwyn, E., and Wandel, J. (1998). *Agroecosystem Health: Analysis and Assessment.* Agroecosystem Health Project, University of Guelph, Ontario, Canada. 114 pp.

Spedding, C.R.W. (1988). *An Introduction to Agricultural Systems.* Elsevier Applied Science, London. 189 pp.

Stangel, P.J. (1995). Nutrient cycling and its importance in sustaining crop-livestock systems in sub-Saharan Africa: an overview. In *Livestock and Sustainable Nutrient Cycling in Mixed Farming Systems of Sub-Saharan Africa.* Vol. 2. Powell, J.M., Fernández-Rivera, S., and Williams, T.O., eds. Addis Ababa: ILCA, pp. 35–57.

Stotz, D. (1983). Production techniques and economics of smallholder livestock production systems in Kenya. In *Farm Management Handbook.* Vol. 4. Jaetzold, R., and Schmidt, H., eds. Nairobi: Ministry of Agriculture, Kenya, in cooperation with the German Agricultural Team (GAT) of the GTZ, pp. 95–106.

Stringer, E. (1999). *Action Research.* 2nd ed. Sage, Newbury Park, CA. 229 pp.

Swindale, L.D. (1992). A research agenda for sustainable agriculture. In: *Impacts of Climate Variations and Sustainable Development in Semi-arid Regions.* Proceedings of International Conference. Magalhaes, A.R., ed. Fortaleza, Brazil: ICID, pp. 123–135.

Tansley, A.G. (1935). The use and abuse of vegetational concepts and terms. *Ecology* 16: 284–307.

Technical Advisory Committee (TAC). (1987). *Sustainable Agricultural Production: Implications for International Agricultural Research.* Technical Advisory Committee of the CGIAR, Rome.

Thompson, J., and Pretty, J.N. (1996). Sustainability indicators and soil conservation. *Journal of Soil and Water Conservation* 51: 265–272.

Tivy, J., and O'Hare, G (1981). *Human Impact on the Ecosystem.* Oliver and Boyd, Edinburgh. 243 pp.

Toledo, V.M. (1990). The ecological rationality of peasant production. In *Agroecology and Small Farm Development.* Altieri, M.A., and Hecht, S.B., eds. Boca Raton, FL: CRC Press, pp. 53–60.

Treitz, W., and Narain, T.M. (1988). Conservation and management of the environment and natural resources in developing countries—Policy implications for ACP states. In *The Changing Dynamics of Global Agriculture.* Seminar/Workshop on Research Policy Implications for National Agricultural Research Systems. Feldafing (Germany), 22–28 September 1988. The Hague, Netherlands: International Service for National Agricultural Research.

van Bruschem, J. (1997). Preface. In *Agroecosystem Health: Proceedings of a Seminar Held in Wageningen, September 26th, 1996.* van Bruschem, J., ed. The Hague, Netherlands: National Council for Agricultural Research, pp. i–ii.

van Voris, P., O'Neil, R.V., Emanuel, W.R., and Shugart, H.H. (1980). Functional complexity and ecosystem stability. *Ecology* 61: 1352–1360.

Walshe, M.J., Grindle, J., Nell, A., and Bachmann, M. (1991). *Dairy Development in Sub-Saharan Africa.* Technical Paper 135. World Bank, Washington, DC. 94 pp.

Waltner-Toews, D. (1991). One ecosystem, one food system: The social and ecological context of food safety strategies. *Journal of Agricultural and Environmental Ethics* 4: 49–59.

Waltner-Toews, D. (1994). Ecosystem health: A framework for implementing sustainability in agriculture. In *Agroecosystem Health with Special Reference to the Consultative Group for International Research (CGIAR): Proceedings of an International Workshop.* Nielsen, N.O., ed. Guelph, Ontario, Canada: Agroecosystem Health Project, Faculty of Environmental Sciences, University of Guelph, pp. 8–23.

Waltner-Toews, D. (1996). Ecosystem health—A framework for implementing sustainability in agriculture. *Bioscience* 46: 686–689.

Waltner-Toews, D., Murray, T., Kay, J.J., Gitau, T., Raez-Luna, E., and McDermott, J.J. (2000). One assumption, two observations, some guiding questions, and a process for the investigation and practice of agroecosystem health. In *Agro-ecosystems, Natural Resources Management and Human Health Related Research in East Africa.* Proceedings of an IDRC-ILRI International Workshop held at ILRI, Addis Ababa, Ethiopia, 11–15 May 1998. Jabbar, M.A., et al., eds. Ottawa: IDRC; Kenya: ILRI, Nairobi, pp. 7–14.

Waltner-Toews, D., and Nielsen, N.O. (1997). *Agroecosystem Health Management and Livestock.* University of Guelph, Ontario, Canada.

Waltner-Toews, D., and Wall, E. (1997). Emergent perplexity: In search of post-normal questions for agroecosystem health. *Social Science and Medicine* 45: 1741–1749.

Winograd, M. (1994). *Environmental Indicators for Latin America and the Caribbean: Toward Land Use Sustainability.* IICA/GTZ Project, World Resources Institute (WRI), Washington, DC. 83 pp.

Winrock International. (1992). *Assessment of Animal Agriculture in Sub-Saharan Africa.* Winrock International Institute for Agricultural Development, Morrilton, AR. 125 pp.

Woomer, P. (1992). What beyond sustainability? Soil fertility research for maize and bean production systems in eastern Africa highlands. In *Proceedings of a Working Group Meeting,* Thika, Kenya, 1–4 September 1992. GAT African Workshop Series No. 21.

World Commission on Environment and Development. (1987). *Our Common Future.* Report of the World Commission on Environment and Development. Oxford University Press, Oxford, UK. 131 pp.

Yudelman, M. (1987). *Prospects for Agricultural Development in Sub-Saharan Africa.* Winrock International Institute for Agricultural Development, Morrilton, AR. 55 pp.

2 Design and Implementation of an Adaptive, Integrated Approach to Health and Sustainability in a Smallholder-Dominated Agroecosystem

2.1 INTRODUCTION

How can knowledge and research be structured to help people make better decisions with regard to managing their agroecosystems? Increasingly, recognition is growing among researchers and development workers that people are part of complex systems (Fitzhugh, 2000). Through various activities, they influence the structure and function of these agricultural and ecological systems to increase the benefits they derive from them, serving—in this way—as the primary managers of the system. The systems, however, consist of extensive, complex, and dynamic interrelationships, such that activity at one point of the system results in complex, sometimes counterintuitive or unpredictable reactions at other spatial or temporal points (Holling, 1986, 1992). Furthermore, the reactions may be lagged in time or difficult to perceive because of the scale at which they occur. Because of these, the consequences of various management strategies are not always easily recognized, making purposeful management of these complex systems difficult.

The concept of health has been found useful in structuring the processes of managing an agroecosystem toward the desired or ideal state (Rapport, 1995; Waltner-Toews and Nielsen, 1995; Haworth et al., 1998). Agroecosystem health is a metaphor that helps to organize knowledge about agroecosystems, structure our evaluative judgments concerning their current state, and reflect them against our hopes for the future so that they (agroecosystems) might be monitored and managed adequately (Haworth et al., 1998). Agroecosystem health management consists of five steps: (1) describing the system of interest; (2) identifying the owners, actors, and customers; (3) setting or naming the goals and objectives of the system; (4) identifying and implementing feasible and desirable changes; and (5) monitoring appropriate indicators, reassessing the situation, and implementing desired changes (Bellamy et al., 1996; Waltner-Toews and Nielsen, 1997).

A systemic description is a model, built using conventional systems theory (Bellamy et al., 1996), the purpose of which is to describe the behavior of the agroecosystem. Agroecosystems, however, can be viewed and interpreted from a variety of nonequivalent perspectives (Waltner-Toews et al., 2000), giving rise to multiple—conflicting or complementing—descriptions (Gitau et al., 1998). Since farmers and communities are the primary managers of the agroecosystem, a managerially useful description is likely to be a synthesis of their perspectives. Colearning tools such as action research (Stringer, 1999) provide means through which such a synthesis can be achieved. By incorporating the primary managers in a collegial participatory process (Biggs, 1989), action research methods provide the framework through which implementation of desired changes and reassessment of the situation can be carried out.

Agroecosystem goals are a reflection of what are considered desirable states for the agroecosystem (Bellamy et al., 1996). According to Haworth et al. (1998), agroecosystem goals can be derived in three ways. The first is a purely subjective process by which expectations for the agroecosystem are decided on a priori based on what is generally regarded as the purpose of the agroecosystem. In the second way the human participants of an agroecosystem form expectations for that agroecosystem. In this sense, system goals are the expected outcomes of transformations that agroecosystem users, owners, or managers would undertake to modify the agroecosystem to optimize the benefits they derive from it. Another way of generating system goals is to study the way the agroecosystem functions, with the selection of system goals a matter of elucidating the goals inherent in the system itself. The three methods represent different points of a continuum; the choice is dependent on the nature of the agroecosystem under study. Whichever way is used to derive system goals, the account of agroecosystem health will consist of a list of goals, a description of the agroecosystem's capacity to meet those expectations, coupled with a list of indicators that enable one to decide how well the system is meeting the expectations (Haworth et al., 1998). Data gathered using these indicators then serve as a basis for refining the system descriptions and management goals (and therefore the indicators themselves) in an iterative, feedback process.

The use of indicators to study complex phenomena is widely accepted (Rapport and Regier, 1980; Odum, 1983; Rapport et al., 1985; Swindale, 1992; Izac and Swift, 1994; Winograd, 1994; van Bruschem, 1997; Aldy et al., 1998; Smit et al., 1998). Their use is complicated by the fact that agroecosystem health is system and scale specific, making the choice of indicators and their interpretation similarly specific. In addition, there is a virtually infinite list of potential indicators. What is needed to implement the broad ideas of health and sustainability is not so much another list of indicators to measure but an integrated framework within which such indicators can be developed and interpreted (Waltner-Toews, 1991). Without a conceptual model that provides a framework for selecting indicators, specifying the data collection and calculation methodologies and a process for synthesizing all the information into a picture of the system, the overall status of the system cannot be assessed (Boyle et al., 2000).

This chapter describes the process used to implement an integrated and adaptive approach to agroecosystem health and sustainability management in a smallholder-dominated tropical highlands agroecosystem. Participatory and action research methods were used to generate system descriptions and to generate local theory (Elden and Levin, 1991) on the management of agroecosystem. Soft system methodologies were used as a tool for creating mutual understanding and for negotiation among the stakeholders so that action plans can be made and implemented. Conventional research methods were used to carry out measurements on selected indicators.

2.2 RESEARCH STRATEGY AND METHODS

Kiambu district, a geopolitically defined region within the Kenyan highlands, was chosen as the study area for two reasons: (1) its proximity to the University of Nairobi (cost considerations) and (2) the fact that it is a district with high agricultural potential and with a preponderance of smallholder farms. The district is relatively more endowed with resources, while agricultural production is more intense than in many other districts. Questions of ecosystem sustainability and health are therefore of greater concern in this district. There are relatively more management options for self-sustenance in Kiambu, therefore providing a suitable venue for testing methods of implementing health and sustainability.

The project involved three groups of actors: (1) communities in six study sites distributed across the district, (2) resource persons comprising extension and technical staff from divisional administrative offices, and (3) researchers. The last group a multidisciplinary team of agronomists, economists, engineers, medical personnel, sociologists, and veterinarians. Additional personnel, including district staff, and experts from governmental and nongovernmental organizations were included when need arose. All people living within each respective intensive study site (ISS) were invited to participate in the village workshops. However, communities decided to elect a committee, referred to as the village AESH committee, to serve as the focal point for action plan implementation and for communication between the community and other actors. There was a resource persons' team in each division of the district. Each team served as the main link between the research team and the communities. A group of six to eight people were selected from a divisional team to be facilitators in participatory workshops organized in study sites within their jurisdiction.

Table 2.1 shows a chronology of the main activities carried out in the project. Initial activities included (1) collection and collation of background information, (2) training of researchers and their assistants in participatory methods, and (3) initial village workshops. Subsequently, the multidisciplinary team attempted to analyze the village systems using loop (influence or spaghetti) diagrams (Puccia and Levins, 1985). It was then proposed that each community should be requested to make similar diagrams to show how they perceived the relationships among factors influencing the health and sustainability of their agroecosystems. A list of potential indicators was then generated and used to carry out a baseline assessment. Concurrently, communities were facilitated to develop their own suite of indicators and to use them to monitor and assess their agroecosystem in a separate process. The researcher-developed suite of indicators was refined using correspondence analysis. The initial

TABLE 2.1
Chronology of Activities Carried Out in a Process to Assess the Health of a Tropical Agroecosystem in the Central Highlands of Kenya

Timescale	Action	Outputs	Actors
April 1997	Secondary data search, collation, and analyses	Hierarchy structure of the Kiambu agroecosystem; choice of scales and sampling strategy	Researchers
May 1997	PAR training	Expertise in PAR methods, visual aids (researchers and assistants)	Research team
June 1997	Sampling study sites	List of study sites, workshop schedules	Researchers and resource persons
July–October 1997	Initial village workshops in the ISS	System descriptions, problem analysis, community action plans	Communities, researchers, and resource persons
September 1997	Multidisciplinary team meeting	System description; problem analysis	Researchers, multidisciplinary team
October–November 1997	Village workshops	Influence diagrams, problem analyses, soft system models	Communities, researchers, resource persons
December 1997	Multidisciplinary team meeting	List of potential research-based indicators	Researchers and resource persons
January–March 1997	Census of land-use units	Typology of land-use units	Communities, resource persons
April 1997	Statistical and system analyses	System attributes, models, and potential indicators	Researchers
May 1998	Multidisciplinary team meeting	A suite of research-based indicators	Researchers, multidisciplinary team
May 1998	Leadership training and intervillage workshop	Understanding of AESH and monitoring and evaluation concepts	Community leaders, researchers, resource persons
June 1998	Multidisciplinary team meeting	Methods for measuring research-based indicators	Researchers, multidisciplinary team
July 1998	Village workshops	Community-driven indicators, AESH training materials	Communities, researchers, and resource persons
August–October 1998	Village workshops	Analyses of community-based indicators data; overall evaluation	Communities
October 1998–January 1999	Development of measurement tools	Questionnaires, semistructured interviews, participatory tools	Researchers, multidisciplinary team

TABLE 2.1 (continued)
Chronology of Activities Carried Out in a Process to Assess the Health of a Tropical Agroecosystem in the Central Highlands of Kenya

Timescale	Action	Outputs	Actors
January–March 1999	Research-based indicator measurement (land use)	Land-use unit-level indicator data	Researchers
April–June 1999	Research-based indicator measurement (study site)	Village-level research-based indicator data	Researchers, resource persons, communities
May 1999	Multidisciplinary team meeting	Approaches to analysis of research-based indicators	Researchers, multidisciplinary team
August 1999–February 2000	Research-based indicator analyses	Refinement of research-based indictors	Researchers
March–November 2000	Village workshops	Monitoring and evaluation using both suites of indicators	Researchers, communities
August 2000	Wrap-up workshop	Overall assessment of the AESH process by the communities	Community leaders, resource persons, multidisciplinary team

AESH, agroecosystem health; ISS, intensive study site; PAR, participatory action research.

phase of the research process was concluded with a wrap-up workshop in which community leaders, resource persons, and some members of the multidisciplinary team discussed the problems, advantages, and disadvantages of the AESH approach. A conceptual framework of the research strategy is summarized in Figure 2.1.

2.2.1 Secondary Data and Holarchical Scales

The purpose of secondary data was to construct a conceptual hierarchical structure of the Kiambu agroecosystem and to identify the scales (in these hierarchies) at which health and sustainability management would best be carried out. Secondary data were used to provide information on the biophysical, economic, and sociopolitical characteristics of the Kiambu agroecosystem. Administrative and topographical maps of the district (Survey of Kenya topology maps 134/3, 134/4, 148/2, 149/1, 148/3, 148/4) provided background data on administrative boundaries, topography, infrastructure, and natural resource endowment. Data on climatic and ecological zonation were derived from the *Farm Management Handbook* (Jaetzold and Schmidt, 1983). Kiambu District Development Plans and reports from various government ministries were used to provide information on existing projects and development plans.

Holarchies were defined from two perspectives: the biophysical and the human activity perspectives. The human activity holarchy followed social, cultural, and

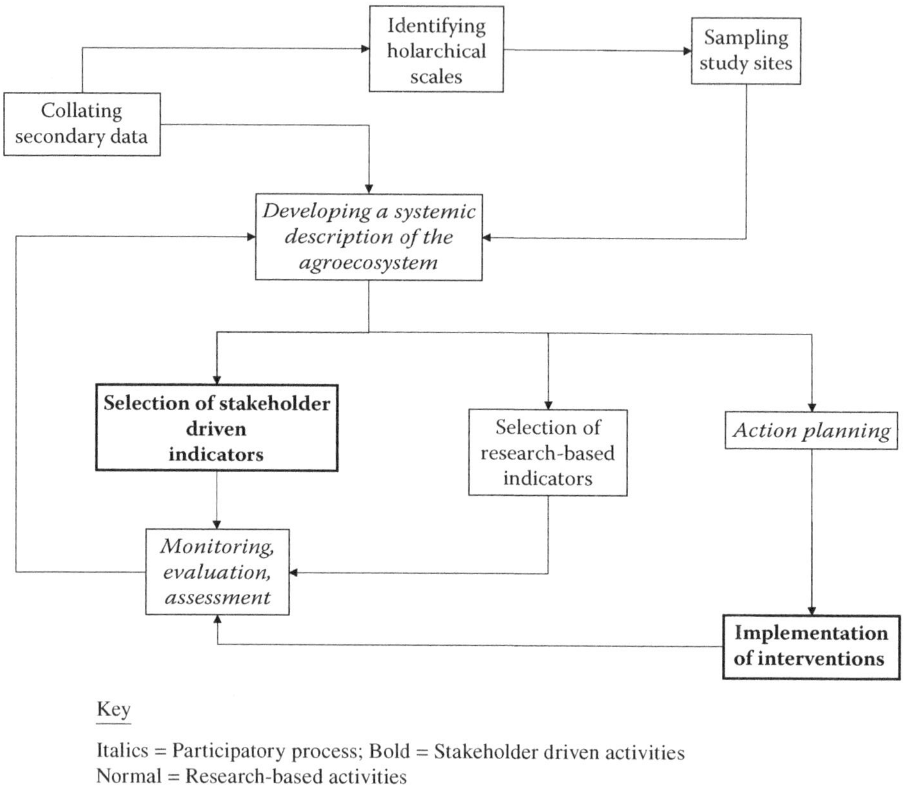

FIGURE 2.1 Flowchart of the research process used to assess and implement health and sustainability of a smallholder-dominated, tropical highlands agroecosystem. See CD for color image.

political boundaries, while the biophysical holarchy was defined mainly by geoclimatic and land use characteristics. The scale at which to carry out the study was decided on based on three considerations. The first was that the health and sustainability of smallholder farms was of most concern in this study. Second, the integration of ecological, economic, and social factors gives rise to emergent properties that are key to the health and sustainability of smallholder farms. Last, the principle stated by Izac and Swift (1994) that to assess sustainability at a given level (n) in the holarchy, both the level above ($n + 1$) and that below ($n - 1$) must also be included in the assessment.

2.2.2 Sampling Study Sites

Once the target hierarchical scales were identified, a sampling strategy for each scale was decided. It was assumed that comparisons among sampling units within each scale as well as an assessment of how they complement and interlink with others would provide sufficient details on the main features of the agroecosystem as a whole. In this study, two sampling units were used. The first were the study sites, corresponding to villages in the human activity holarchy and catchments in the biophysical. The second sampling units were the land-use units, roughly corresponding to farms in the biophysical holarchy and to households or homesteads in the human activity holarchy. *Land-use units* were defined as parcels of land separated by formal boundaries shown on ordinance survey maps. *Households* were defined as people living under the same roof or sharing food from the same kitchen. Homesteads were groups of households within the same land-use unit, with no formal boundaries between them.

The Kiambu agroecosystem was stratified into regions based on the holarchical scales in the human activity system. A stratified purposive sampling protocol was used to select study sites. The criteria for selection were preponderance of smallholder farmers (favored if more) and the number of development agencies (favored if less). This was done by the resource persons using a participatory scoring matrix. In total, 12 sites (2 in each main holarchical division) were selected. Six of the study sites (one in each division) were labeled "intensive" (ISS) and the others "extensive" (extensive study sites, ESSs) using a random protocol. The ISSs were those study sites in which health and sustainability interventions were instituted.

2.2.3 Systemic Description and Action Planning

The objective was to obtain a systemic description of the agroecosystem based on the perspectives of the people living in the ISSs. The process commenced with participatory workshops in each of the six ISSs. The local language, Gikuyu, was used as the main language of communication between community groups and the research team. These workshops had three objectives: (1) a systemic description of the agroecosystem, (2) participatory problem analysis, and (3) community action planning. Data on (1) boundaries, (2) natural resources, (3) institutional structure, (4) historical background, (5) social structure, (6) farming system characteristics, (7) economic and climatic trends, (8) human health, (9) constraints to health and well-being of the residents, and (10) their coping strategies were gathered, analyzed, and presented using a variety of participatory tools. The workshops culminated with participatory problem analysis and action planning. Details of the methods used are presented in Chapter 3.

One-day workshops were held in each of the ISSs 4–6 weeks later. In these, participants (the village committee and at least one representative from each household/homestead) were asked to make similar influence diagrams based on their perception of these relationships. The resulting diagrams were analyzed using graph theory (Bang-Jensen and Gutin, 2001), qualitative methods (Puccia and Levins, 1985), and pulse process modeling (Perry, 1983). Details of these analyses are presented in Chapter 4.

Descriptions and pictures of the problematic situations identified in each of the ISSs (holons) were developed using approaches described by Checkland and Scholes (1990). Relationships among various institutions and interest groups were explored and depicted in rich pictures (Checkland, 1979a). In addition, root definitions (Checkland, 1979b) were made for each intervention in the community action plans. These definitions, descriptions, pictures, and models were used in two ways: (1) to identify both the sources and the types of conflicting or competing perspectives, goals, and action plans; and (2) as tools for generating a common understanding of a problem situation and for negotiating some degree of consensus on goals and plans. These are discussed in detail in Chapter 5.

To determine the types and characteristics of the units comprising the penultimate layer of the study sites, a census of all land-use units within each of the six ISSs was carried out. In this census (Appendix 1) details on the (1) characteristics of the owners and managers, (2) types and quantities of resources available, (3) types of enterprises carried out within them, (4) constraints to productivity, (5) goals and objectives, and (6) productivity were sought. Gini coefficients and Lorenz curves as described by Casley and Lury (1982) were used to explore the distribution of resources among the land-use units. Gini coefficients were calculated as $(T1 - T2)/10{,}000$, where $T1$ is the sum of the cross products of cumulative percentage of land-use units and lagged cumulative percentage of the resource. $T2$ is the sum of the cross products of lagged cumulative percentage of land-use units and cumulative percentage of the resource. The Gini coefficient lies between 0 (absolute equality) and 1 (absolute inequality). If two distributions are compared, the one with a larger coefficient is more unequal, but this depends on the shape of the Lorenz curves. If the distribution with a smaller coefficient lies entirely within the other, then the conclusion about relative inequality is unequivocal. If the curves cross each other, then the inequalities differ only over parts of the range of these distributions.

2.2.4 Indicators

Two methods were used to generate two suites of indicators. Communities, through a participatory process facilitated by the researchers, developed the first set suite. Researchers and the multidisciplinary team developed the second suite using descriptions given by the communities in the initial workshop and in the loop diagrams. Details of the process and methods used are presented in Chapter 6, Section 6.2.

2.2.4.1 Community-Driven Indicators

The objective for the community-driven indicators was to develop a suite of indicators that the communities can use to assess the health and sustainability of their agroecosystem. The indicators were developed in two stages. First, discussions were initiated among communities during leadership training programs with regard to the AESH concept and the ideas of monitoring and evaluation. Three-day workshops were then held in each of the six villages; the indicators were developed at these workshops. Participatory tools such as focus group discussions, scoring matrices, and trend analyses were used to identify, rank, and then categorize indicators. Further details on the participatory methods used are provided in Chapter 3.

2.2.4.2 Selection of Research-Based Indicators

For research-based indicators, the objective was to develop a suite of indicators for use by researchers and policymakers. It was assumed that this suite of indicators would be complementary to the community-driven suite. Indicators were defined as variables that reflect (1) changes in key system attributes or (2) changes in the degree of risk or potential of the system. Indicators were selected based on the ease of measurement and interpretation, validity, cost-effectiveness, and usefulness of the information gathered to researchers and policymakers. Further details are provided in Chapter 6.

2.2.5 Monitoring, Evaluation, and Assessment

2.2.5.1 Community-Based Assessments

Participatory monitoring, evaluation, and assessments were carried out in ISSs only. This was based on the assumption that self-monitoring provides communities with information that is crucial to the successful management of the agroecosystem. It was also assumed that self-evaluation would create a sense of ownership of the process by the communities, and that this would enhance their participation, thereby increasing the sustainability of the process. By understanding how communities evaluated information gathered using indicators, it was hoped that researchers would gain insight on how indicators can be analyzed to be of use in practical decision making.

Monitoring was taken to mean the evaluation of indicators on a daily or weekly basis to provide information on the progress of specific community activities. Such information would be used for short-term management and decision making. *Evaluation* was defined as a review of goals and objectives against achievements. This would occur after completion of specific activities or attainment of predefined milestones. Evaluation could also be done regularly after a defined period to evaluate progress toward overall community goals. *Assessment* was defined as an overall review of the agroecosystem status in terms of health and sustainability using selected indicators.

2.2.5.2 Research-Based Assessments

Research-based assessments were carried out in all 12 study sites in February 1998 and again in February 1999. Empirical data on research-based indicators were gathered using both conventional research methods and participatory tools. A questionnaire (Appendix 2) was developed and applied to each of the land-use units in each of the 12 study sites. Process and methods used are discussed in Chapter 6.

2.2.6 Implementation of Interventions

The objectives were to reinforce the communities' capacity for collective remedial action. The underlying assumption was that health and sustainability depended on the communities' ability to design appropriate remedial actions and to implement them successfully. Community participation was seen to be the key to the sustainability of the process. Two types of interventions were therefore envisaged. The first was to impart analytical, management, and participatory skills to the communities to enhance their capacity for problem identification and analyses, consensus building, conflict resolution, action planning, monitoring, evaluation, and assessment. The second type of intervention was to provide expertise and support geared toward facilitating communities in the implementation their action plans.

2.2.6.1 Community Training

Training programs were organized in each of the six ISSs and at the district level. Village AESH committee members, some opinion leaders, and 6–10 people from the ISSs were trained on participatory approaches, management methods, community mobilization, gender issues, community-based leadership, action planning, monitoring, and evaluation. Experts from the various disciplines were invited to conduct training in each of the specialized areas. Focus group discussions were held after each topic. The experts then addressed specific issues arising from these discussions. Leaders in each of the ISSs were encouraged to hold monthly village meetings to discuss, in a participatory manner, their agroecosystem sustainability and health concerns.

2.2.6.2 Community-Based Development Interventions

Leaders in each of the ISSs were provided with copies of the action plans developed in the participatory workshops. The research team facilitated meetings among the community leaders in each village and between them and other institutions to discuss the implementation of action plans and to institute measures for better management of their agroecosystem. The leaders were expected to initiate participatory processes to develop activity schedules, delegate duties, monitor progress, and evaluate the progress of individual projects.

The implementation of the action plans was the responsibility of the communities. In addition, the communities were expected to supply all the resources needed to carry out the required interventions. The role of the research team was to identify experts, resource persons, or institutions that the communities might need for successful implementation of a project. If the resources needed for a project were more than the communities could generate from within, information and skills (e.g., proposal writing) for seeking support from the government, nongovernmental organizations, and other development agencies were provided. However, communities were requested to show how such a project would be sustained after the donor support ceased.

2.3 RESULTS

Figure 2.2 shows the relative size and location of Kiambu district. Change in altitude (in units of 200 m starting from sea level) is also shown to illustrate the location and extent of the highlands. The geographical distribution of the study sites within Kiambu district and the relative size of the divisions are illustrated in Figure 2.3. The boundaries of the newly created Tigoni Division were yet to be properly documented by the time of this study.

Communities in all selected study sites agreed to participate. Community participation was high, with 75% to 100% of the households and homesteads represented in all the participatory workshops held in the study sites. The concept of AESH was well understood by the stakeholders as evidenced by use of the health language and concepts during the participatory workshops.

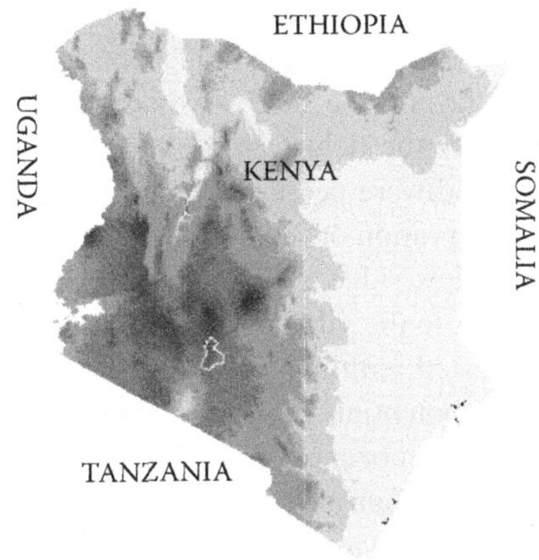

FIGURE 2.2 Map of Kenya showing the location and relative size of Kiambu district and the highlands. See CD for color image.

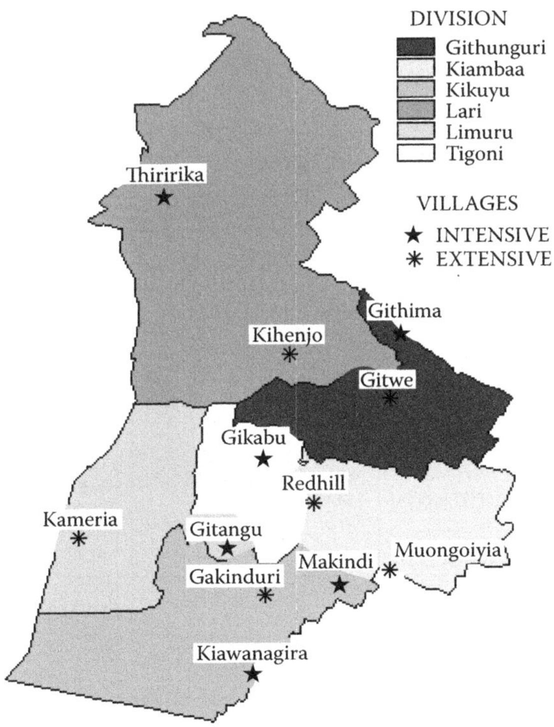

FIGURE 2.3 Map of Kiambu showing the administrative divisions and the locations of intensive and extensive study sites. See CD for color image.

2.3.1 Holarchical Scales

The biophysical holarchy is best described in terms of five layers (Figure 2.4). The innermost or smallest layer—the field—was defined mostly by management characteristics. The layer after the field was the farm. Farms were defined mostly by land-use characteristics and were perceived as nested within catchments (a term commonly used by soil conservation officers in the district). The latter were defined mostly by topographical (valley, ridge, plain, etc.) characteristics. Catchments corresponded, in many instances, to the villages defined in the human activity holarchy. Catchments were seen as nested within agroecozones as described by Jaetzold and Schmidt (1983). Agricultural potential, vegetation, and geologic and climatic factors defined the boundaries of agroecozones. Kiambu is within the central highlands geoclimatic zone and comprises four major agroecozones (Figure 2.4).

The human activity holarchy was confluent with the administrative zoning of Kiambu district. The district is divided into six administrative zones called divisions (Limuru, Kikuyu, Lari, Tigoni, Githunguri, and Kiambaa). Each division is further subdivided into several locations, which are in turn divided into sublocations (Figure 2.4). The latter is the lowest formal administrative unit. According to the key informants and administrative officials, each sublocation may consist of one to four villages with informal boundaries, but consisting of groups of people who work

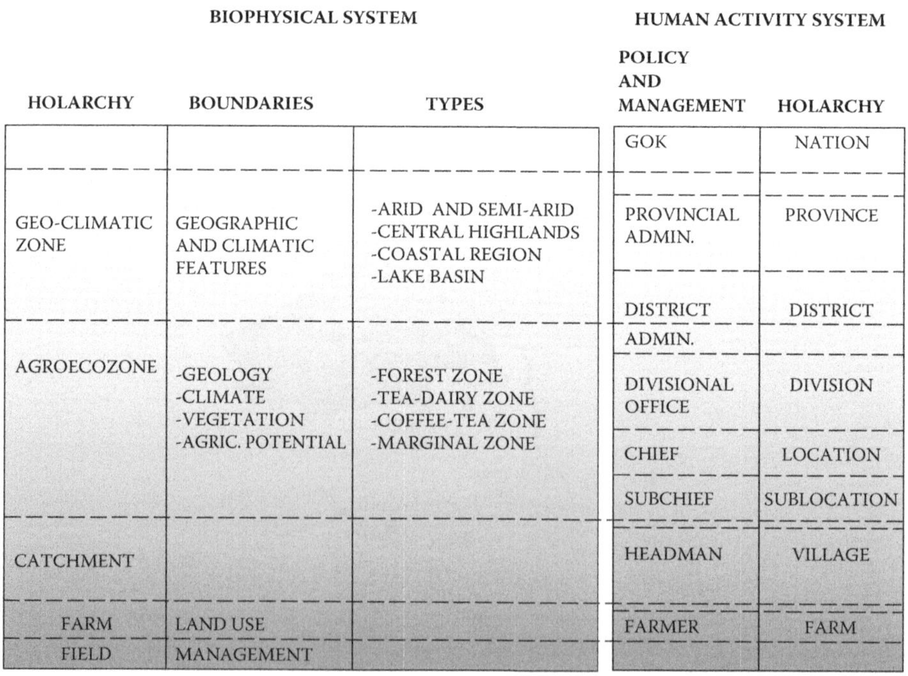

FIGURE 2.4 An illustration of the holarchical structure of the Kiambu agroecosystem from both the biophysical and the human activity perspectives. GOK, Government of Kenya. See CD for color image.

together as a unit. Village boundaries are defined through different criteria, including topographical features. It is possible for villages to lie across administrative boundaries. Secondary data listing villages or describing their boundaries could not be found. Within homesteads and households, systems of management define several farm enterprises, comprising the lowest rung of the human activity holarchy. For health and sustainability management of the Kiambu agroecosystem, the village level and the household level were selected as the most appropriate scales for AESH management.

2.3.2 Study Sites

Participatory mapping confirmed the presence of villages as a layer nested within the sublocation in the human activity holarchy. Sociocultural factors were more important in defining the boundaries of the villages. Communities regarded themselves as belonging to one of these villages, with various sociocultural institutions organized and functioning at this level.

Githima village has boundaries that are confluent with administrative ones. The village is described as the area under the administrative jurisdiction of the assistant chief. Another identity factor was the use of two coffee-processing factories and three tea-buying centers in the area. People settled in the village prior to 1952, clearing an indigenous wattle tree forest.

Gitangu village derives its identity partly from its historical background (area inhabited by three subclans) and from administrative boundaries (area under an assistant chief). The area is an indigenous forest occupied by hunter-gatherers. Settlement by the current tribe began before the arrival of Europeans. The three subclans (*Mbari-ya-igi*, *Mbari-ya-Gichamu,* and *Mbari-ya-Ngoru*) derive from the three people who first settled in the area.

Deriving its identity from its geophysical location (a swampy valley bounded by roads and railway) and its sociocultural history, Kiawamagira is inhabited by descendants of squatters in the Church Missionary Society Mission in Thogoto. Elders claimed that during the land demarcation process, those squatters who were not considered favorably by the mission were allocated land in the valley.

Mahindi village lies on a ridge between two streams and is inhabited by members of the *Kihara* subclan. The name of the village refers to the elephant skeletons found on the ridge. Settlement started in the 1950s. The boundaries of Gikabu-na-Buti village of Tigoni Division are socioeconomic. The village adjoins another, and both are sandwiched within two vast tea estates. The land was part of one of the tea estates and was sold to a cooperative of its laborers. Settlement began in 1972. Itungi village consists of 4-acre land parcels, while Gikabu-na-Buti village consists entirely of half-acre plots, thereby creating a socioeconomic subdivision within what seems to be a single village. During the initial mapping exercise, participants indicated that they were one village. In subsequent meetings, it was revealed that the two are disparate with very few interactions between them. The sixth village, Thiririka, was described as the area under the administrative jurisdiction of an assistant chief. This was part of Kinale forest until 1989, when land was allocated to settle squatters from various forests in the district.

2.3.3 Systemic Description

Gitau (1997) provided a detailed description of the information gathered during the initial village workshops. This includes descriptions of natural resources, historical background, social structure, typology of farms, trends, human health, seasonal calendars, felt needs, and coping strategies by communities living in the six ISSs.

2.3.3.1 Demographic Features

Table 2.2 gives a summary of key demographic features of the six ISSs based on a census of land-use units. The Githima study site had the highest number of land-use units (229), followed by Gitangu. Kiawamagira and Mahindi had the fewest (41 and 40, respectively). The mean acreage per land-use unit was highest in Thiririka (3.5 acres), followed by Mahindi (2.7 acres) and Githima (2.3 acres). Kiawamagira and Gikabu had the least (1.8 and 1.9, respectively). In terms of total size, Thiririka is the largest in land size, covering approximately 3 km^2 and having several publicly owned parcels of land. Mahindi and Kiawamagira are the smallest in size, covering approximately 0.5 km^2 each. There were areas in Kiawamagira left as public land due to swamping.

In all villages, there were land-use units that consisted of more than one household (Table 2.2). These were more common in Githima (23) and Gitangu (19) and least common in Mahindi and Kiawamagira (1 and 6, respectively). Nearly half (43.9%) of the households in Kiawamagira were female headed. The majority of the households in Gikabu (63.9%) and Kiawamagira (53.7%) were managed by females. The majority of households in Mahindi (67.5%) and Gikabu (57.8%) had off-farm income. The average number of people per household was highest in Thiririka (8 persons), followed by Mahindi, and the fewest people were in Githima households (5.6 persons). Mahindi had the highest number (2.5) of people with off-farm employment per household, followed by Gikabu (1.5) and Kiawamagira (1.4), while Githima had the fewest (0.3).

2.3.3.2 Geoclimatic Features

According to the agroecological classification by Jaetzold and Schmidt (1983), Thiririka village lies in the forest reserve zone (Upper Highlands; UH0 in Figure 2.5) as shown in Figure 2.5. Githima village lies in the coffee-tea zone (upper midlands; UM1). Mahindi and Kiawamagira villages lie in the marginal coffee zone (upper midlands; UM3). The other two villages are on the lower highlands (LH) zones: Gitangu in the wheat-maize-pyrethrum zone (LH2) and Gikabu in the tea-dairy zone (LH1).

2.3.3.3 Resource Use and Distribution

Off-farm employment, small ruminants, and income from various farming enterprises were the most unevenly distributed. Gini coefficients were 0.72 for off-farm employment, 0.28 for population, 0.41 for farm land, 0.43 for cattle, 0.69 for sheep and goats, 0.64 for income from cash crops, 0.53 for income from food crops, and

TABLE 2.2
Summary of Key Demographic Features Based on a 1997 Census of Land-Use Units in the Intensive Study Site, Kiambu District, Kenya

	Githima	Gitangu	Mahindi	Thiririka	Kiawamagira	Gikabu
Division	Githunguri	Limuru	Kiambaa	Lari	Kikuyu	Tigoni
Approximate size of village (km^2)	2	2	0.5	3	0.5	1
Number of land-use units	229	224	40	188	41	83
Mean acreage per unit	2.3 ± 0.17	2.1 ± 0.12	2.7 ± 0.34	3.5 ± 0.14	1.8 ± 0.21	1.9 ± 0.19
Units with more than one household	23	19	1	9	6	15
Number of households	304	296	41	230	62	147
Proportion of female-headed households	22.7%	18.8%	30.0%	17.0%	43.9%	27.7%
Proportion of female managed households	31.9%	46.4%	50.0%	32.4%	53.7%	63.9%
Proportion of households with off-farm income	14.8%	37.5%	67.5%	29.8%	36.6%	57.8%
Mean number of people per household	5.6 ± 0.25	6.1 ± 0.22	7.8 ± 0.6	8.0 ± 0.35	7.3 ± 1.0	7.0 ± 0.36
Mean off-farm employed per household	0.3 ± 0.06	0.8 ± 0.10	2.5 ± 0.4	0.6 ± 0.09	1.4 ± 0.35	1.5 ± 0.20
Mean number going to school per household	2.3 ± 0.12	2.7+0.15	1.7 ± 0.28	2.8 ± 0.15	2.24 ± 0.31	2.5 ± 0.24

0.54 for income from livestock. Population was evenly distributed in all six villages, as were farmland and cattle (Table 2.3).

In Mahindi, all eight resources considered were equitably distributed. In Kiawamagira, only off-farm employment was markedly uneven, while in Gikabu it was only income from food crops. Off-farm employment was most unevenly distributed in Githima (Figure 2.6), while sheep and goats were unevenly distributed by

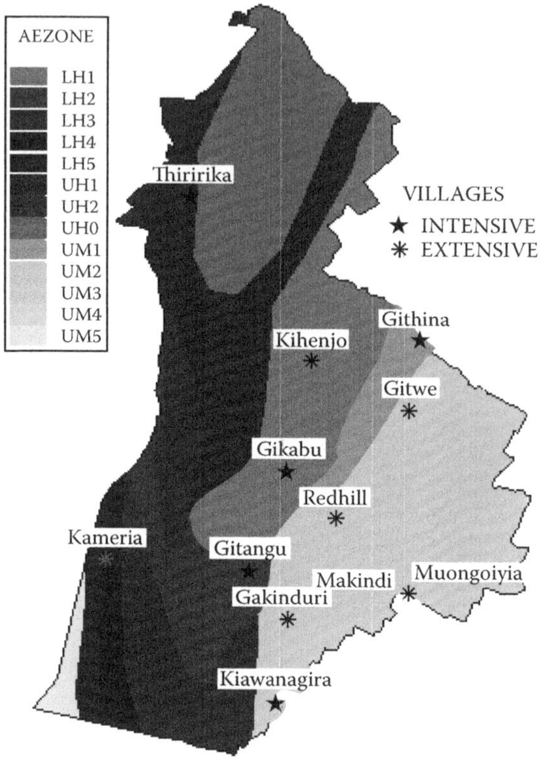

FIGURE 2.5 Map of Kiambu showing the distribution of study sites by agroecozones as described by Jaetzold and Schmidt (1983). LH, lower highlands; UH, upper highlands; UM, upper midlands. See CD for color image.

about the same magnitude in Githima, Gitangu, and Gikabu (Figure 2.7). Income from food crops was the most inequitable in Githima (Figure 2.8), in contrast to Thiririka, where income from cash crops (coffee and tea) was the most inequitable (Figure 2.9).

Figure 2.10 shows the proportion of land under various crops in the six ISSs based on the 1997 census of land-use units. For each village, the proportion of land under each enterprise was calculated as the average of the per farm proportion. Most (36.44%) of the farmland in Githima village was allocated to coffee. In Kiawamagira village, most (49.70%) of the land was under food crops. In Thiririka, nearly 50% of farmland was left fallow or as pasture. Gitangu village had the highest (17.39%) proportion of land allocation to fodder (mostly Napier) among the six villages (Gikabu 9.18%; Githima 6.52%; Kiawamagira 7.72%; Mahindi 12.58%; and Thiririka 0.52%). In addition, the land allocated to noncrop activities (other) was proportionately bigger because of space used for housing livestock and paddocks. Horticultural crops had the biggest proportion of farmland in Thiririka village.

Figure 2.11 shows the location of public medical facilities in Kiambu district relative to both ISSs and ESSs. Of the ISSs, only Thiririka and Mahindi were close to a public health facility (within 1-km radius of the village; closest facility for other

TABLE 2.3
Equity, Measured by the Gini Coefficient, of Key Resources in the Intensive Study Site, Kiambu District, 1997[a]

Resource	Githima	Gitangu	Mahindi	Thiririka	Kiawamagira	Gikabu
Off-farm employment	0.72	0.61	0.43	0.68	0.53	0.35
Population	0.26	0.25	0.18	0.29	0.14	0.22
Farm land	0.44	0.35	0.23	0.27	0.28	0.40
Cattle	0.44	0.39	0.38	0.39	0.40	0.34
Sheep and goats	0.70	0.66	0.45	0.54	0.34	0.61
Income from cash crops	0.54	0.09	0.00	0.42	0.00	0.32
Income from food crops	0.56	0.47	0.31	0.48	0.26	0.52
Income from livestock	0.44	0.40	0.15	0.66	0.20	0.43

[a] Gini coefficients were calculated using the method described by Casley and Lury (1982) as (T1-T2)/10,000 where T1 is the sum of the cross-products of cumulative percentage of land-use units and lagged cumulative percentage of the resource. T2 is the sum of the cross-product of lagged cumulative percentage of land-use units and cumulative percentage of the resource.

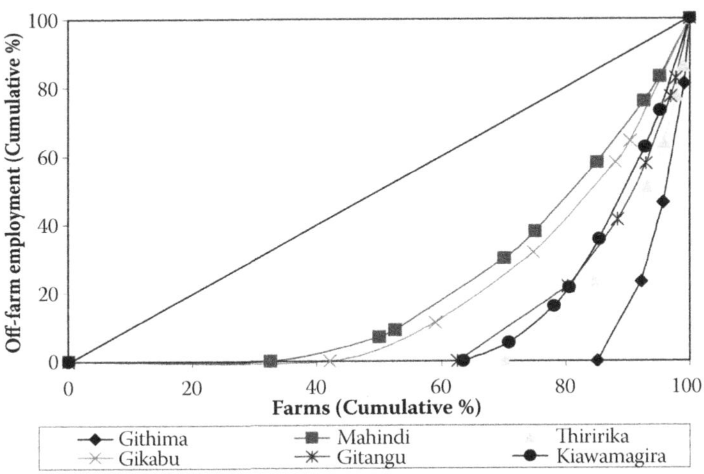

FIGURE 2.6 Lorenz curve showing the distribution of off-farm employment in the intensive study site, Kiambu District, Kenya, 1997. See CD for color image.

villages was 10 km). Private health facilities, however, were available within Gikabu, Kiawamagira, and Githima villages.

The distribution of water supply schemes in the district relative to the study sites and major urban centers is shown in Figure 2.12. Among the ISSs, only Githima, Kiawamagira, and Gitangu were located within areas covered by a water supply scheme. At the time of this study, Komothai water scheme (covering Githima study

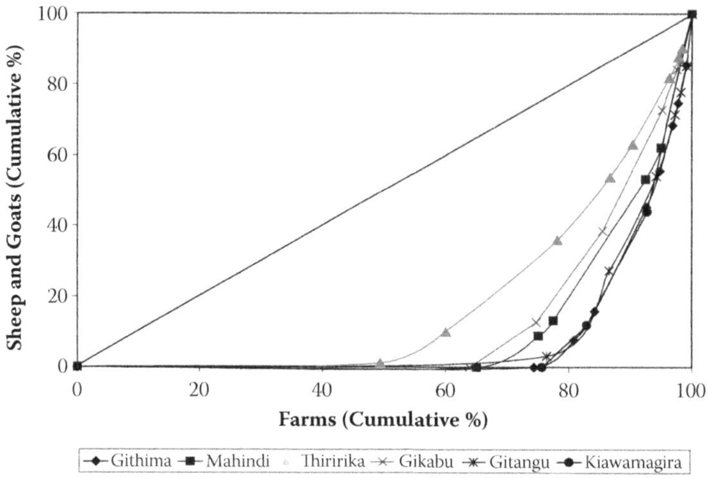

FIGURE 2.7 Lorenz curve showing the distribution of sheep and goats in all intensive study sites, Kiambu District, Kenya, 1997. See CD for color image.

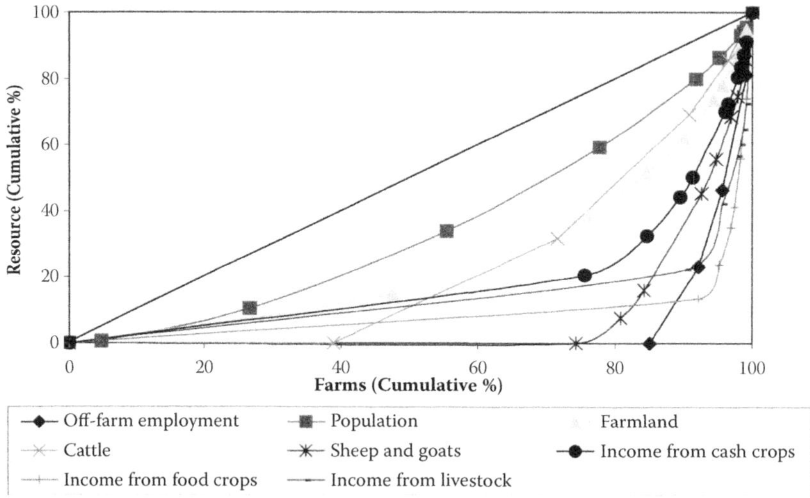

FIGURE 2.8 Lorenz curves of eight key resources in Githima intensive study site, Kiambu District, Kenya, 1997. See CD for color image.

site) was not operational, reportedly because of silting of the main dam. The Ngecha water scheme, covering Gitangu village, was also not operational following theft of the pumping equipment. In both cases, the water supply infrastructure (pipes and tanks) was still present but in a state of disrepair.

2.3.3.4 Agriculture

The main agricultural products in Githima were coffee and tea, while in Gitangu, Mahindi, Thiririka, and Kikuyu they were dairy and vegetables. In Gikabu, the

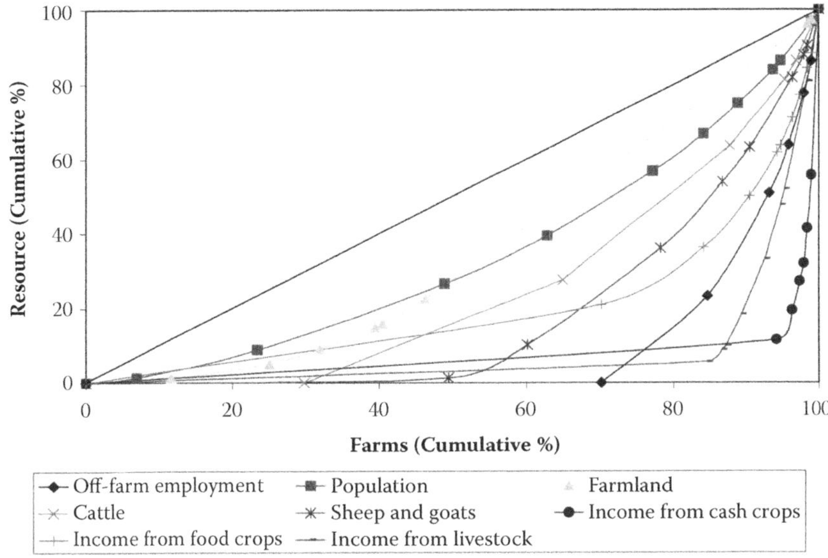

FIGURE 2.9 Lorenz curves of eight key resources in Thiririka intensive study site, Kiambu District, Kenya, 1997. See CD for color image.

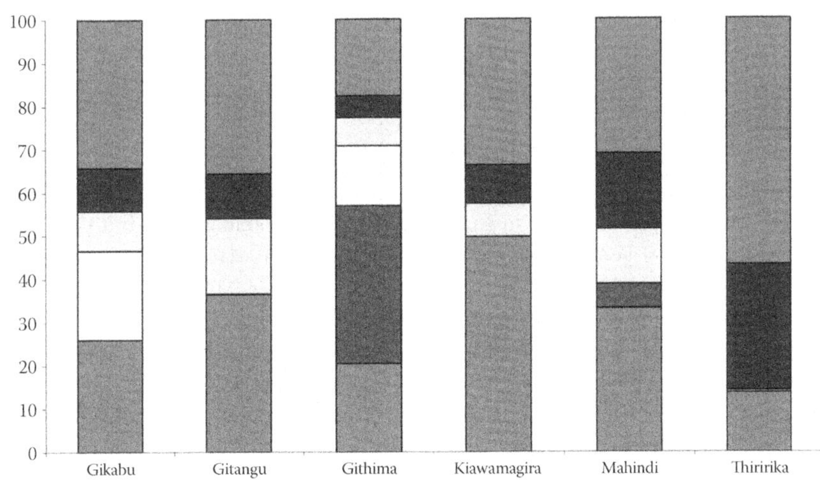

FIGURE 2.10 Allocation of land resource to various crops in the intensive study site (ISS) Kiambu district, 1997. For each village, the average per-farm acreage of a crop was expressed as a percentage of the average farm size in that village. See CD for color image.

main products were tea and dairy. Thus, only Githima, Thiririka, and Gikabu had agricultural activities confluent with their agroecological classification. The other three villages were mainly focusing on dairy and horticultural vegetable production, irrespective of their agroecological classification. Little or no pyrethrum was being

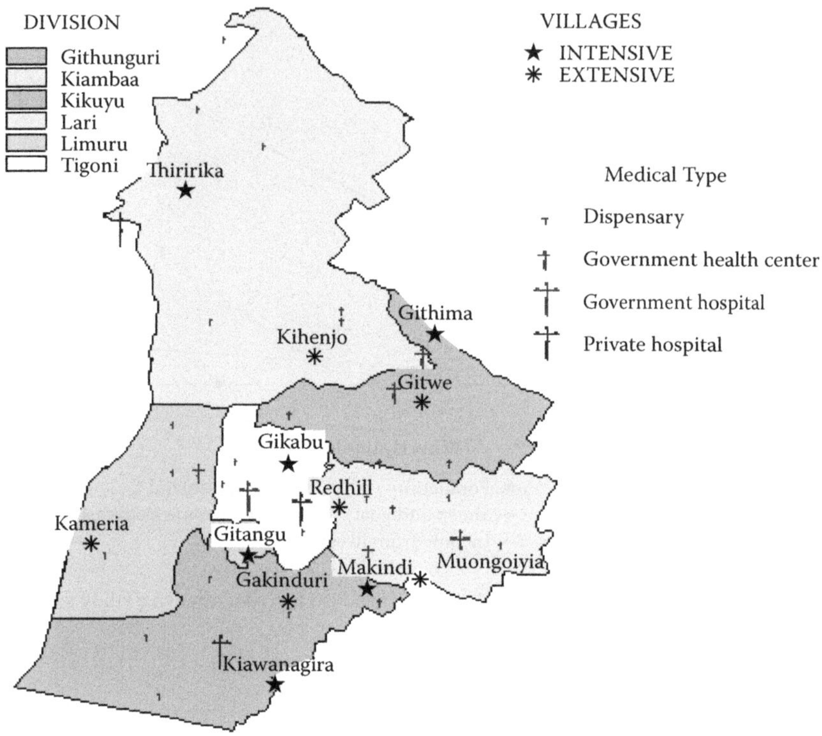

FIGURE 2.11 Map of Kiambu showing the distribution of medical facilities. See CD for color image.

produced in Gitangu village, and there was no coffee production at all in Kiawamagira. There were a few farmers in Mahindi village who had coffee, but they had not had a harvest for 10 years. The reason given was that coffee was not profitable to produce in this village.

A comparison of the relative importance of the three main farm enterprises (cash crop, food crop, and livestock), based on their contribution to the annual farm income, is shown in Figure 2.13. Proportions were computed for each farm and then averaged for each village. Most (84.88%) of the farm income in Githima village came from traditional cash crops (coffee and tea), while that in Mahindi (62.67%) came from the sale of surplus food crops (maize, beans, potatoes, kale). In Thiririka, farm income was mainly (57.09%) from sale of horticultural produce, especially vegetables. Farm income in Gikabu was balanced among tea, dairy, and food crops (especially kale). Livestock were the major (77.87%) contributors to farm income in Gitangu village and contributed an important proportion (35.42%) of it in Kiawamagira. Annual farm income was highest in Githima village, followed by Gitangu, while it was lowest in Mahindi, followed by Kiawamagira (Figure 2.14). In contrast, income per acre of land was highest in Githima, followed by Kiawamagira. It was lowest in Mahindi and Thiririka villages.

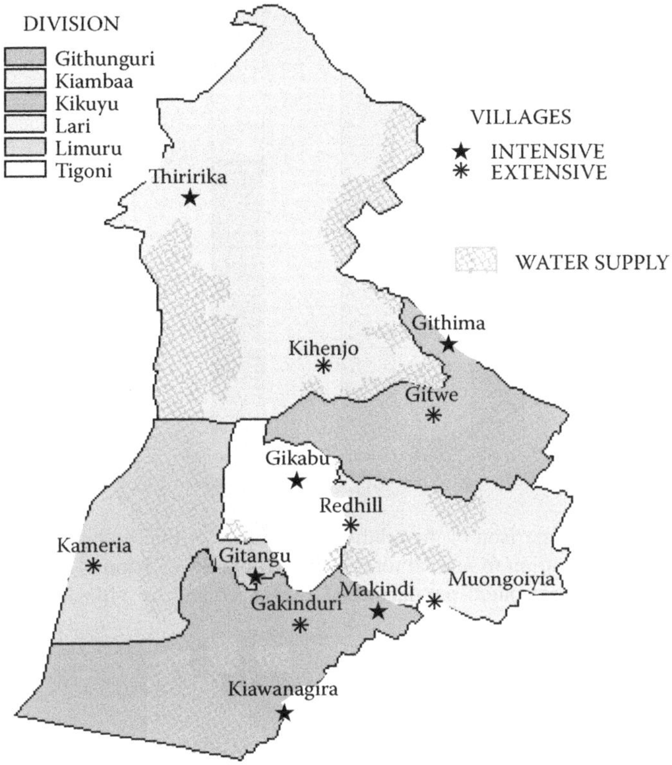

FIGURE 2.12 Map of Kiambu District showing the coverage of different water supply schemes. See CD for color image.

Table 2.4 shows the agricultural products, inputs, and crop and animal diseases in each of the six villages. Githima, Thiririka, and Gikabu reported herbicides as one of the major external inputs for their villages in terms of quantity and expenditure. Fertilizer was considered an important input in all villages, but it was relied on heavily only in Githima and Thiririka villages. Farmers classified farm enterprises as livestock (*Mahiü*), food crops (*irio*), and nonfood (cash) crops. The livestock enterprise was further classified as commercial poultry, cattle, and small ruminants (*mbüri*) and local chicken. Food crops were further distinguished by whether they were mainly for consumption within the farm (subsistence) or for sale. Food crops grown for subsistence were mostly maize, beans, potatoes, and peas. Food crops grown mainly for the market were vegetables, especially kale (horticulture). Different cash crop enterprises (coffee, tea, and pyrethrum) were always specified and considered separate. The farmers' choice of enterprises was governed mostly by the tradition in the area, experience of the manager, availability of resources, availability of market for the produce, and the potential yield of the enterprise.

Horticulture was considered the most important in terms of income in all villages except Githima. The main crop produced was kale (*Sukumawiki*), which has

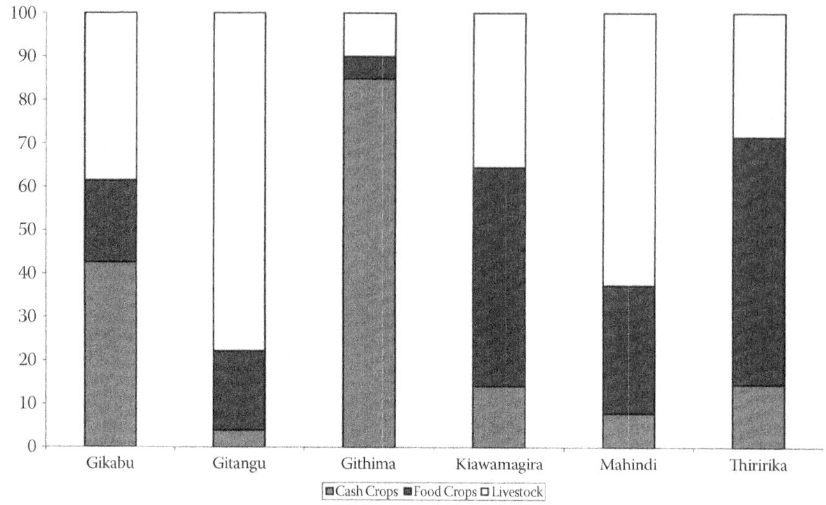

FIGURE 2.13 A comparison of the relative importance of the three main farm enterprises based on their contribution to annual total farm income. Proportion of annual income from each enterprise was computed per farm and then averaged by village. See CD for color image.

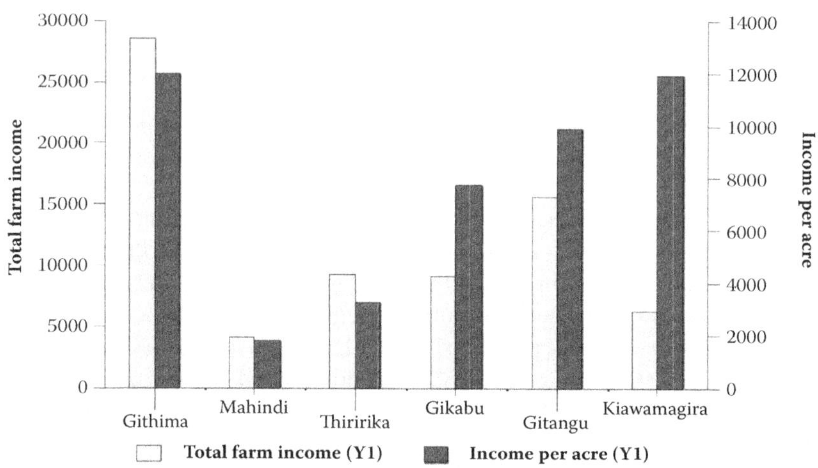

FIGURE 2.14 Comparison of the annual farm incomes in the six intensive study sites. Total farm income was calculated as all sales for 1997 minus all farm-related costs except casual and household labor. Income per acre was calculated as the annual farm income divided by the farm size (in acres). See CD for color image.

a ready market in Nairobi. The capital outlay was minimal, and return to labor was considered high. The limitations were seen to be transportation and soil fertility. Yield was high during the rainy season, but the villages are not accessible during this time, so most of the produce goes to waste. Disease and pests are also an important

TABLE 2.4
The Main Products, Inputs, Crop Pests, and Livestock Diseases Reported by Farmers in Farm Censuses Carried Out in the Intensive Study Sites

	Githima	Mahindi	Thiririka	Gikabu	Gitangu	Kiawamagira
Food crops (mainly for home consumption)	Maize Beans Potatoes Arrowroots Yams Bananas	Maize Beans Potatoes Arrow roots Bananas	Maize Beans	Maize Beans Potatoes Arrowroot Peas	Maize Beans Potatoes Peas Banana	Maize Beans Potatoes Arrowroots Bananas
Cash crops	Coffee Tea			Tea	Pyrethrum	
Horticulture	Kale Tomatoes	Kale Celery Flowers Sugar cane	Potatoes Kale Carrots Peas Onions Pears	Kale Pears Onion Plums	Kale Flowers Oranges Carrots Avocado Coriander	Mangoes Avocados
Livestock	Cattle Sheep Goats	Cattle Sheep Goats Poultry	Cattle Sheep	Cattle Sheep	Poultry Cattle Sheep Donkey	Cattle Sheep Goats
Fodder	Napier	Napier	Napier Oats Pasture	Napier	Napier Crop residue	Napier
External inputs	Fertilizer Manure Herbicides Feed	Manure Seeds Fertilizer Feed	Herbicides Fertilizer Seeds Feed Rent tractor	Herbicides Fertilizer Feed	Feed Seeds Manure	Feeds Manure Herbicides Fertilizer
Labor	Casual Family	Family	Casuals Family	Permanent Casual Family	Permanent Casual Family	Family
Crop and livestock diseases and pests	Coffee BD FMD Milk fever Pneumonia	Moles Hedgehogs Blight Aphids Mastitis Worms ECF	Blight Aphids Frost Foot rot Mastitis Pneumonia	Moles Hedgehogs Stock borer Bacterial wilt Mastitis Milk fever ECF	Blight Moles Hedgehogs Gumboro Ndigana	Bacterial wilt Blight Stock borer Weevils
Off-farm activities		Employment Hawking	Business	Employment Business	Employment	Business Employment

BD, Berry Disease; FMD, Foot and Mouth Disease; ECF, East Coast Fever consideration in kale production. Farmers were conscious of the environmental and health impacts of chemical control.

2.3.3.5 Agroecosystem Health Goals

2.3.3.5.1 Participatory Method

Details of problems and concerns—as identified and prioritized by participants in the village workshops—are given in Chapter 3. Concerns common to all ISSs were availability of water for domestic use, poor roads, poor human health, and absence of health care facilities. Only one village (Kiawamagira) had access to piped water, and even then, the water was available for one half-day per week. Roads were mainly loose surface, becoming impassable during the wet season. Due to the hilly terrain of the Kiambu agroecosystem, flooding and gully formation are the biggest causes of poor road condition. Among the agriculture-related problems were lack of artificial insemination services, low crop yields, poor soil productivity, lack of markets for produce, lack of extension services, and crop and animal diseases.

According to the participants, the main limitation to crop production in these villages was land size, but climate and market (price) were also important. Limitation to dairy production was seen to be mainly capital and feed-related constraints. Food crop production was reported as limited mainly by soil fertility, which in turn is a consequence mainly of soil erosion and depletion of nutrients. Poultry production was reported as severely limited not only by diseases, especially Gumboro and Newcastle, but also by market for eggs and meat. In terms of livestock, dairy cattle were given a higher preference to small ruminants and poultry since the milk market is available, and the returns were said to be higher. In all ISSs, consensus on needs and goals was achieved. Committees of local participants were selected to oversee the implementation of the action plans.

2.3.3.5.2 Survey Method

In the land-use unit survey, 35.3% of the respondents reported lack of extension services as a constraint to productivity (Table 2.5). In contrast, 33.8% of the farmers reported soil infertility as a constraint; land size was a constraint for 14.4% of the

TABLE 2.5
Constraints to Productivity as Reported by Respondents in a Survey of Land-Use Units in the Six Intensive Survey Sites

	Small Farm Size	Soil Infertility	Inadequate Extension	Lack of Labor	Lack of Capital	Flooding
Githima (% of 229 units)	23.1	70.3	48.9	3.1	14.8	0.0
Mahindi (% of 40 units)	22.5	52.5	47.5	10.0	0.0	0.0
Thiririka (% of 188 units)	7.4	27.1	31.4	1.1	3.2	14.9
Gikabu (% of 83 units)	15.7	21.7	28.9	4.8	8.4	0.0
Gitangu (% of 224 units)	7.6	4.9	25.0	17.9	0.9	0.0
Kiawamagira (% of 41 units)	24.4	24.4	34.1	2.4	0.0	7.3
Overall (% of 805 units)	14.4	33.8	35.3	2.2	6.1	3.9

respondents. Soil infertility was a more common problem in Githima (23.1%) and Mahindi (22.5%) villages. Githima (14.8%) village had more respondents reporting lack of capital than in other villages. Flooding and waterlogging was reported only in Thiririka and Kiawamagira villages.

Nearly all the respondents (96.3%) indicated that they would like to improve farm productivity, whether by starting new enterprises or improving existing ones (Table 2.6). The majority of respondents not willing to improve farm productivity were in Gikabu (11/30) and Thiririka (12/30) villages. Most of the respondents reported that they would prefer to improve livestock (mainly dairy) production and horticulture. In Githima village, most respondents reported that they preferred to enhance cash crop (tea or coffee) production compared to all the other options.

2.3.4 Health and Sustainability Assessment

Communities understood the concepts of health and health indicators and accepted the notion of using indicators to evaluate the status of their agroecosystem. They appeared to regard the approach not as an innovation, but as a revisiting and modernization of traditional methods of agroecosystem management.

Communities in the ISSs opted to carry out agroecosystem evaluations and assessments jointly with other ISSs. The communities initiated intervillage collaboration because they felt that participants from other study sites provided additional useful criticism and suggestions compared to those by the researchers and extension agents.

2.3.5 Implementation of Interventions

At the end of the initial village workshops, all communities expressed a profound demand for action to ameliorate the problems identified. Formation of the village committees was seen as evidence of their desire to implement the action plans. Five of the six villages proceeded with implementation of the action plans immediately after the workshops, mostly without further contact or consultation with the research team. In nearly all the cases, this led to some degree of frustration on the part of the communities as they were ill prepared in terms of organization and community leadership to carry out many of the tasks. However, there were some successes, and failure and frustration did not deter most of the communities from continuing to try. Further details are provided in Chapter 3. Details on the methods used to facilitate planning and implementation of action plans by the communities are provided in Chapter 4.

2.4 DISCUSSION

2.4.1 Holarchical Scale

There were two reasons why the village was selected as the target level for this study. Foremost of these is that at the village/catchment level, ecological, economic, and social factors are integrated, resulting in unique emergent properties. Second, tradeoffs among farms within a village are essential factors in the sustainability of agriculture in the entire Kiambu ecosystem. The land-use level was selected because it is

TABLE 2.6
Stated Goals of Respondents in Census of All Land-Use Units in the Six Intensive Study Sites, Kiambu District, 1997

	Improve Cash Crop	Improve Livestock	Improve Food Crops	Improve Horticulture	Improve Business	Other Improvements	No Improvements
Githima (% of 229 farms)	81.2	59.0	59.8	71.6	4.8	4.4	1.3
Mahindi (% of 40 farms)	7.5	45.0	50.0	40.0	2.5	0.0	0.0
Thiririka (% of 188 farms)	6.9	50.5	35.6	34.0	12.2	2.1	6.4
Gikabu (% of 83 farms)	30.1	55.4	36.1	20.5	22.9	16.9	12.5
Gitangu (% of 224 farms)	22.8	64.3	34.4	58.9	3.6	1.3	0.0
Kiawamagira (% of 41 farms)	12.2	63.4	29.3	41.2	22.0	9.8	9.8
Overall (% of 805 farms)	35.2	57.6	42.6	50.9	8.8	4.4	3.7

the basic agricultural unit and forms the penultimate layer of the village/catchment in the Kiambu holarchy. Trade-offs and other interrelationships among land-use units have a significant impact on the farm's health and sustainability. Land-use units corresponded to two conceptually different levels in the human activity holarchy (homestead, household), resulting in difficulties in sampling and interpretation. Homesteads were an aggregation of semiautonomous households with no formal boundaries between then. These resulted from the cultural practice by which sons—married or single—build their houses around their father's houses. It is not clear to what extent the reemergence of this phenomenon is a response to diminishing land sizes per household and to what extent it represents some form of intergenerational inequity.

Although the existence of the village as level of organization in the human activity holarchy was expected, the level of social and cultural integration found was surprising. Many formal and informal associations and organizations functioned at this level. Communities were acutely aware of the boundaries of their village, and certain sociocultural activities were exclusively undertaken within these boundaries. Resources outside these boundaries were considered not available to the community no matter how close.

2.4.2 Systemic Description

The descriptions provided by the communities in the initial workshops and in their loop diagrams indicated that communities were acutely aware, in a systemic way, of the biophysical and social economic factors important in determining the health and sustainability of their agroecosystems and hence their livelihoods. Communities were unencumbered by disciplinary training and provided systemic descriptions that were detailed enough to make them useful but not bogged down in detail.

Villages—both extensive and intensive—closer to Nairobi were smaller both in terms of area covered and in the number of households (Table 2.2). The structure of the level penultimate to the village also varied by distance to Nairobi, with the closer villages having fewer land-use units under the management of more than one household. Furthermore, these villages had more people with off-farm employment, while a bigger proportion of the land was allocated to subsistence food crops. In addition, agricultural production had evolved away from the traditional modes for which these areas were considered most suitable. The effects of distance from Nairobi appear to be related to urbanization, availability of off-farm employment, and an accessible market for food crops. The difference in farm income between Kiawamagira and Mahindi—the villages closest to Nairobi—was most likely because off-farm employment was the main source of income in Mahindi.

Agriculture seemed to be dominated by coffee, tea, dairy, and kale production. An interesting finding was that the average per acre income in Kiawamagira, where vegetable production is predominant, was similar to that of Githima village, where traditional cash crops (coffee and tea) were produced. In contrast, the farm income per acre in Thiririka village, in which vegetable production was predominant, was less than half that of Kiawamagira, with the total farm income almost equal. This seems to indicate an increased intensification as land sizes become smaller.

The villages were densely populated. Based on the numbers presented in Table 2.2, the population density ranged between 600 and 1,030 persons per square kilometer. However, because the villages were selected in a purposive process, this may not be reflective of the density of the entire Kiambu agroecosystem. It is, however, likely to be the case in those areas where smallholders were predominant. The high proportion of female-managed households reflects the tradition of men seeking off-farm employment in the urban areas and rural-based industries while women remain in the home and manage the farms.

2.4.3 The Agroecosystem Health Approach

This work demonstrates that a holistic approach to investigating AESH and beginning to implement sustainable processes for AESH improvement is feasible even in complex field situations. More important, however, is that communities were able to use the concept of health to discuss and model approaches to improve their livelihoods. The approach provided a simple, yet highly specialized, language—understood by the communities, researchers, extension agents, development agents, and policymakers—for discussing issues relating the health and sustainability of agroecosystems and for structuring the process through which remedial actions can be taken.

A unique feature in this process was that community, researchers, and development agents played complementary roles. Using Biggs's (1989) framework, which described the relationship between researchers and communities in terms of the extent to which local opinion and practice is given recognition, this process would fall into the category referred to as collegial. The community's role was crucial in understanding the system and posing the key questions of interest. Through participatory problem analysis and action planning, the community's informal research-and-development system was actively encouraged.

Researchers and resource persons played an important role of facilitating the implementation of the action plans (e.g., offering technical advice, providing research activities to answer key community questions, facilitating contacts with outside agencies, writing proposals to investors, and providing leadership training), but the leading role was left to the communities. Research questions of broader interest, such as social analysis of communities, indicator development and application, and determinants of sustainability were investigated with community input and collaboration in both the design and analysis. The main output from the interaction between the two (communities and researchers) was a synergy that augmented both the communities' and the researchers' ability to first detect and then investigate and act on AESH concerns.

The main difficulties in the approach relate to its time horizon, broad perspective, and location specificity. As the process is open-ended, only its initiation and early development fit into a standard project time frame. Longer-term issues, such as assessing sustainability, require longer-term mechanisms of support. The holistic view adopted in this process, while essential to establishing the crucial context for decisions, means that sometimes there was a lack of decision-making focus. Last, from a research perspective, it is not yet clear how generalizable the lessons learned are. In our view, the process is potentially transferable, but this requires further study. The main limitation to this is that the process is highly dependent on the communities' capacity for collective action. Conceivably, absence of collective action would cripple the communities' ability to participate, plan, take action, and reflect on the outcomes of such actions. The effectiveness of methods such as soft systems methodology in rebuilding communities' social capital and collective action remains to be fully tested.

The AESH framework as applied in this study is remarkably similar to the sustainable livelihoods approach. Similarities include the holistic and systems approaches, focus on the communities as partners in the process, community participation, and seeking sustainable transformations of human activity systems to improve the well-being of the people. AESH is a metaphor to structure how human beings should think about their activities—social and economic—and their implication on the biophysical world not only to improve their well-being but also to conserve the natural resource base on which their survival depends. The sustainable livelihood approach, by contrast, seeks to develop an understanding of the factors that lie behind people's choice of livelihood strategy, reinforce the positive aspects, and mitigate the negative influences (DFID [Department for International Development], 2000). It presupposes the existence of a consensus on what is positive and negative among the outsiders and insiders, while its primary purpose is to enhance the outsiders' understanding of the choices of the insider so that the outsiders can design and implement better development interventions.

Agroecosystem health proposes a shift toward communities as the primary managers of agroecosystems, with a colearning process in which insiders are experts on the problem situation, while outsiders are experts on the methods. Together, they function to create local theory to be used in a collegial process to improve the livelihoods of communities and the health and sustainability of agroecosystems as a whole. The AESH framework therefore adds to traditional methods of integrated community development by incorporating systems theory and practice, action research concepts, and participatory as well as conventional research methods to address potentially multiple and varied community-driven concerns. This allows a more structured approach for addressing complex societal issues such as equity, gender, and leadership roles.

In planning and implementation, communities are willing—and often need—to enlist the advice and support of "outsiders" for addressing priority concerns. Specific research questions may be posed (e.g., regarding water quality) or technical advice requested (e.g., about specifications for water distribution or road construction). We have also found that communities are receptive to learning from the experiences of other communities, which is useful for providing both practical tips and motivation. All of these are critical to the process of encouraging communities toward healthy and more sustainable husbandry of agroecosystems and underscore the potential of the AESH approach.

2.4.4 Health and Sustainability Assessment

Although the methods and strategies used in this study provided important results, it is difficult to assess how well they predict the health and sustainability of the

Kiambu agroecosystems given the short span of the project. However, the fact that communities, policymakers, and researchers are using information resulting from these assessments to make decisions suggests that the approach succeeded, to some degree, in integrating health and sustainability concerns in the decision-making systems of Kiambu district. In addition, all remedial actions taken as a result of these assessments seem to be a move toward sustainability and better AESH. Further monitoring will be needed before an assessment of time and cost-effectiveness of the process can be meaningfully assessed.

REFERENCES

Aldy, J.E., Hrubovcak, J., and Vasavada, U. (1998). The role of technology in sustaining agriculture and the environment. *Ecological Economics* 26: 81–86.

Bang-Jensen, J., and Gutin, G. (2001). *Digraphs: Theory, Algorithms and Applications.* Springer-Verlag, New York. 754 pp.

Bellamy, J.E.C., Waltner-Toews, D., and Nielsen, N.O. (1996). *Agroecosystem Health—A Fuzzy Systems Approach.* Discussion Paper 30. Agroecosystem Health Project, University of Guelph, Ontario, Canada. 20 pp.

Biggs, S.D. (1989). *Resource-Poor Farmer Participation in Research: A Synthesis of Experiences from Nine Agricultural Research Systems.* OFCOR Comparative Study Paper No. 3. ISNAR, The Hague, Netherlands. 37 pp.

Boyle, M., Kay, J.J., and Pond, B. (2000). Monitoring in support of policy: an adaptive ecosystem approach. In *Encyclopaedia of Global Environmental Change.* Munn, R., ed. New York: Wiley, pp. 116–137.

Casley, D.J., and Lury, D.A. (1982). *Monitoring and Evaluation of Agricultural and Rural Development Projects.* Johns Hopkins University Press, London. 145 pp.

Checkland, P., and Scholes, J. (1990). *Soft Systems Methodology in Action.* Wiley, Chichester, U.K. 329 pp.

Checkland, P.B. (1979a). Techniques in soft systems practice. Part 1: Systems diagrams—some tentative guidelines. *Journal of Applied Systems Analysis* 6: 33–39.

Checkland, P.B. (1979b). Techniques in soft systems practice. Part 2: Building conceptual models. *Journal of Applied Systems Analysis* 6: 41–49.

DFID (Department for International Development). (2000). *Sustainable Livelihoods Guidance Sheets.* DFID, London, U.K. 41 pp.

Elden, M., and Levin, M. (1991). Cogenerative learning: Bringing participation into action research. In *Participatory Action Research.* Whyte, W.F., ed. Newbury Park, CA: Sage, pp. 127–142.

Fitzhugh, H. (2000). Foreword. In *Agroecosystems, Natural Resource Management and Human Health Related Research in East Africa.* Proceedings of an IDRC-ILRI international workshop held at ILRI, Addis Ababa, Ethiopia, 11–15 May, 1998. Jabbar, M.A., Peden, D.G., Saleem, M.A.M., and Li Pun, H. IDRC (Ottawa); Kenya: ILRI, Nairobi.

Gitau T. (1997). *Report of participatory action-research workshops held in six villages of Kiambu District.* The agroecosystem Health Project, Department of Public Health, Pharmacology and Toxicology. University of Nairobi, Nairobi, Kenya.

Gitau, T., McDermott, J.J., Waltner-Toews, D., Gathuma, J.M., Kang'ethe, E.K., Kimani, V.W., Kilungo, J.K., Muni, R.K., Mwangi, J.M., and Otieno, G.O. (1998). Agroecosystem health: A framework for ecologically sustainable development. In *Proceeding of the 15th International Association of Farming Systems Research and Extension in Pretoria, South Africa.*

Haworth, L., Brunk, C., Jennex, D, and Aral, S. (1998). A dual-perspective model of agroecosystem health: System functions and system goals. *Journal of Agricultural and Environmental Ethics* 10: 127–152.

Holling, C.S. (1986). The resilience of terrestrial ecosystems, local surprise and global change. In *Sustainable Development of the Biosphere*. Clark, W.C., and Munn, R.E., eds. Cambridge, U.K.: Cambridge University Press, pp. 292–317.

Holling, C.S. (1992). Cross-scale morphology, geometry, and dynamics of ecosystems. *Ecological Monographs* 62: 447–502.

Izac, A.-M.N., and Swift, M.J. (1994). On agricultural sustainability and its measurement in small-scale farming in sub-Saharan Africa. *Ecological Economics* 11: 105–125.

Jaetzold, R., and Schmidt, H. (1983). *Farm Management Handbook of Kenya: Volume IIB—Natural Conditions and Farm Management Information, Central Kenya*. Ministry of Agriculture, Kenya, in cooperation with the German Agricultural Team (GAT) of the GTZ, Nairobi.

Odum, H.T. (1983). *Systems Ecology: An Introduction*. Wiley, Chichester, U.K. 644 pp.

Perry, E.L. (1983). A pulse process model of athletic financing. In *Modules in Applied Mathematics*. Lucas, W.F., Roberts, F.S., and Thrall, R.M., eds. New York: Springer-Verlag, pp. 141–164.

Puccia, C.J., and Levins, R. (1985). *Qualitative Modeling of Complex Systems*. Harvard University Press, Cambridge, MA. 259 pp.

Rapport, D.J., and Regier, H.A. (1980). An ecological approach to environmental information. *Ambio* 9: 22–27.

Rapport, D.J. (1995). Ecosystem health: What's in a name? *Ecosystem Health* 1: 1–2.

Rapport, D.J., Regier, H.A., and Hutchinson, T.C. (1985). Ecosystem behaviour under stress. *American Naturalist* 125: 617–640.

Smit, B., Waltner-Toews, D., Rapport, D., Wall, E., Wichert, G., Gwyn, E., and Wandel, J. (1998). *Agroecosystem Health: Analysis and Assessment*. Agroecosystem Health Project, University of Guelph, Ontario, Canada. 114 pp.

Stringer, E (1999). *Action Research*. 2nd ed. Sage, Newbury Park, CA. 229 pp.

Swindale, L.D. (1992). A research agenda for sustainable agriculture. In *Impacts of Climate Variations and Sustainable Development in Semi-arid Regions*. Proceedings of International Conference. Magalhaes, A.R., ed. Fortaleza, Brazil: ICID, pp. 123–135.

van Bruschem, J. (1997). Preface. In *Agroecosystem Health: Proceedings of a Seminar Held in Wageningen, September 26th, 1996*. van Bruschem, J., ed. The Hague, Netherlands: National Council for Agricultural Research, pp. i–ii.

Waltner-Toews, D. (1991). One ecosystem, one food system: The social and ecological context of food safety strategies. *Journal of Agricultural and Environmental Ethics* 4: 49–59.

Waltner-Toews, D., Murray, T., Kay, J.J., Gitau, T., Raez-Luna, E., and McDermott, J.J. (2000). One assumption, two observations, some guiding questions, and a process for the investigation and practice of agroecosystem health. In *Agro-ecosystems, Natural Resources Management and Human Health Related Research in East Africa*. Proceedings of an IDRC-ILRI international workshop held at ILRI, Addis Ababa, Ethiopia, 11–15 May 1998. Jabbar, M.A., et al., eds. Ottawa: IDRC; Kenya: ILRI, Nairobi.

Waltner-Toews, D., and Nielsen, N.O. (1997). *Agroecosystem Health Management and Livestock*. University of Guelph, Ontario, Canada.

Waltner-Toews, D., and Nielsen, O. (1995). *Assessing Agroecosystem Health*. Agroecosystem Health Project Discussion Paper 23. Agroecosystem Health Project, University of Guelph, Ontario, Canada.

Winograd, M. (1994). *Environmental Indicators for Latin America and the Caribbean: Toward Land Use Sustainability*. IICA/GTZ Project, World Resources Institute (WRI), Washington, DC [s.d.]. 83 pp.

3 Community Participation and the Integration of Agroecosystem Health and Sustainability Concerns into Practical Decision Making

3.1 INTRODUCTION

Agroecosystem health and sustainability are value-based and change-oriented concepts. Both require that issues concerning people, power, and praxis be explicitly addressed. Active participation of communities in agroecosystem health and sustainability assessment and implementation is based on four key principles. The first is that those who experience a socioeconomic phenomenon are the most qualified to describe and investigate it (DePoy et al., 1999). The second is based on the proposition by Lewin that causal inferences about human activity systems are more likely to be valid when the human beings in question participate in building and testing them (Argyris and Schon, 1991). The Freirian theme that poor people can and should be enabled to conduct an analysis of their own reality (Freire, 1968) is another predicate for the inclusion of communities in the process.

Another reason for a participatory approach is that agroecosystem health and sustainability are not objectively verifiable states of a hard system, which means that actions geared toward some long-term plans—but based on current evaluations of health and sustainability—are likely to become less relevant as the system evolves over time and space. Emphasis should shift to iterative planning, implementation, and reflection coupled with continuous monitoring and regular evaluation of progress toward the long-term goals. These processes of planning, action, and reflection should be structured in such a way that they are self-perpetuating, confluent with the local context, and operational within the local decision-making process. The only practical way of achieving this is by enhancing the capacity of communities in the agroecosystem to monitor, plan, and implement their own health and sustainability programs.

In the recent past, several techniques for the systematic involvement of communities in research and development processes have evolved in various dimensions (Chambers, 1994; Jiggins, 1995). Although this has been gainful in many ways, the various evolutionary lines have retained similar (but conceptually disparate) terminologies, such as participatory research, participatory action research (PAR), participatory appraisal, activist participatory research, and participatory rural appraisal (PRA)—causing a lot of confusion. In addition, there are differences within each of these dimensions in the way methods are applied in practice. The common tenet among these approaches is the concept of community participation. Most practitioners apply the term *community participation* to mean some form of interaction between local people and outsiders in which the former play a role in identifying, implementing, or even controlling research or development activities (Catley, 1999). However, the degree and nature of involvement differs widely among various groups of practitioners, resulting in more variations in methods.

Among the most widely used and more homogeneous of the participatory methods are PRA and PAR. PRA has been defined as an intensive, systematic, but semi-structured learning experience carried out in a community by a multidisciplinary team that includes community members (Theis and Grady, 1991). It has also been described as an approach for learning about rural life and conditions from, with, and by rural people (Chambers, 1994). PRA is intended to enable local people to conduct their own analysis and often to plan and take action (Webber and Ison, 1995) in collaboration with outsiders. In contrast, PAR is defined as a form of action research in which professional researchers operate as collaborators with members of organizations in studying and transforming those organizations (Greenwood et al., 1993). It incorporates the principle of iterative cycles of planning, analysis, and action into a collaborative process between researchers and communities (Whyte, 1991). PAR is a way of learning how to explain a particular social world by working with the people who live in it to construct, test, and improve theories about it so they can better control it (Elden and Levin, 1991). An important distinction between the two approaches is that, operationally, PRA is a single, initial phase of interaction between communities and outsiders (Webber and Ison, 1995), while PAR is a structured, ideally unending process of action and evaluation by communities in collaboration with outsiders. The visual representations and analysis by local people (such as mapping; scoring and ranking with seeds, stones, or sticks; group discussions and presentations; and diagramming) are similar between PAR and PRA and among other participatory approaches.

The development of PAR was fueled mostly by industry in the 1980s; loss of competitiveness led managers in industry to shift emphasis toward worker participation in solving problems in productivity and costs (Whyte, 1991). The term *action research* was coined in the 1940s by Kurt Lewin, an American sociologist working on a range of community projects concerning integration and social justice in areas such as housing and employment (Webb, 1996). It refers to a collaborative inquiry by a group of people into a shared problem, issue, or concern for which they feel responsible and accountable and that they seek to solve through teamwork (Zuber-Skerritt, 1996). It attempts to solve problems, issues, or concerns by following a cyclical process of (1) strategic planning, (2) action, (3) evaluation, and (4) revising the plan (Zuber-Skerritt, 1996).

In action research, *collaboration* means that everyone's point of view will be taken (with equal weight) as a contribution to resources for understanding the situation (Winter, 1996). The analysis proceeds to assemble the differences between viewpoints and the contradictions within each of them. In this way, many of the claims made from each viewpoint are translated into questions, allowing for a range of alternatives to be suggested when previously particular interpretations would have been taken for granted. The goal of this process is to generate a set of ideas that have been interpersonally negotiated (Winter, 1996). A form of action research, termed *emancipatory action research* (Carr and Kemmis, 1986), aims at not only resolving the primary concern of the participants but also changing the system itself and those conditions that impede desired improvement. It aims at empowering and increasing the ability of participants to create grounded theory (Glaser and Strauss, 1967), which is a theory developed on the basis of experience and practice and aimed at facilitating the solution of complex problems in different situations.

It is important in both theory and practice to distinguish between the various forms of action research (Whyte, 1991). Elden and Levin (1991) conceived the participatory form of action research as consisting of "insiders" (local participants) and "outsiders" (the professional researchers) collaborating to cocreate "local theory" that the participants test out by acting on it. They defined *local theory* as the most direct, simple, and elegant context-bound explanation of cause-and-effect relations in a given situation that makes sense to those with the most local experience. According to this definition, a local theory is situation specific. It is generated by insiders in dialogue with outsiders using general knowledge and the rules of scientific inquiry and expressed using everyday language and meanings.

The initial framework of what develops into local theory is a description of how individual members of an organization perceive the problem situation. Insiders have their own ideas or models for attributing meaning and explanations to the world they experience. Since they (the insiders) spend most of their lives in the situation of interest, they know more about it and have more ways of making sense of their world than would be possible for an outsider to appreciate without in some way becoming an insider. Thus, insiders are experts in the specifics of the situation. They know from personal experience how things work and how the elements are connected to each other and about values, attitudes, and local culture, factors among those that interact to create the subsisting situation.

Insiders are primarily concerned about theories of their own particular situation—those that would facilitate the solution of practical problems and achievement of personal and organizational goals. Their theories, however, are (in most cases) not systematically tested, and their knowledge is highly individual, tacit, and unreflected on (Elden and Levin, 1991). Outsiders have what is missing: (1) training in systematic inquiry and analysis and expertise in (2) designing and carrying out research and (3) recognizing patterns and creating new knowledge that is less context specific. The second framework that contributes to local theory comes from the application of these principles to generate data about the problem situation and carry out relevant analyses.

In the context of agroecosystem health and sustainability, PAR provides a means through which communities can be involved as collaborators. Specifically, PAR provides the methodological background for collaborating with the communities to (1) generate a systemic description of the agroecosystem, (2) build consensus on management goals for the agroecosystem, (3) plan and undertake remedial action,

(4) develop suites of indicators of health and sustainability, and (5) monitor progress, assess health and sustainability, and evaluate the status of the agroecosystem. This chapter describes how PAR was used to develop a suite of health and sustainability indicators and to implement some actions to address agroecosystem health and sustainability concerns in the tropical highland agroecosystem.

3.2 PROCESS AND METHODS

The process involved three groups of actors: (1) communities in six study sites distributed across Kiambu district, (2) resource persons (extension and technical staff from divisional administrative offices), and (3) researchers. The researchers were a multidisciplinary team of agronomists, economists, engineers, medical personnel, sociologists, and veterinarians. Additional personnel, including district staff, and experts from governmental and nongovernmental organizations were included when need arose.

All the people living within the study sites were invited to participate in most of the village PAR workshops. Communities decided to elect a contact group (committee) to serve as the focal point for communication between the community and other actors in the project. Election to the committee was stratified based on gender, age, and other study-site-specific criteria such as clan and wealth ranking. There was a resource persons' team in each division of the district, serving as the main link between the research team and the study sites in their divisions. From these teams, groups of six to eight people were selected to serve as facilitators in PAR workshops in their division.

Based on the scheme developed by Elden and Levin (1991), the resource persons and the research team comprised the outsiders, while communities in the study sites were the insiders. Similarly, the objective of the process was described as developing grounded, local theory on assessment and improvement of agroecosystem health and sustainability. The process through which the study sites were selected is described in Chapter 2.

3.2.1 Community Identities

The approach used in this study assumed that there would be identifiable communities in each of the study sites. A *community* was defined as a group of local people sharing similar interests (Ison, 1993; Webber and Ison, 1995) and capable of undertaking some degree of collective action. As described by Burkey (1993), it was expected that conflicts of interest, contradictions, and differences in perspectives would exist among different groups within a community. Further, it was expected that a cooperative context within which people have sufficient security to speak and act publicly (Chataway, 1997) might not exist.

The existence, identity, and characteristics of communities in the study sites were determined through initial participatory workshops held in each of the study sites. The geophysical boundaries of the study sites were then altered to be as confluent as possible with those of the communities. To elucidate the interests, composition, and structure of the various groups in the community, root definitions (Checkland

and Scholes, 1990) were constructed for institutions, associations, organizations, social groups, and cooperatives with membership from the study site. Focus groups designed along group boundaries were used to obtain group specific data. These were compared to data generated in presentations to account for instances when participants are unable or unwilling to speak or act in the presence of others. Where complex and messy problem situations (such as lack of a cooperative context, people unable to speak or act publicly, unbridgable conflicts of interest, irreconcilable contradictions, and differences in perspectives) existed, soft systems methodology (Checkland and Scholes, 1990) was adopted. The use of soft systems methodology is described in Chapter 5.

3.2.2 Initial Contact with Communities in the Study Sites

The initial contact with communities in the intensive study sites (ISSs) was through public meetings. First, an awareness campaign was carried out in the selected areas through administration officials (chiefs and assistant chiefs) and in churches and markets using posters and presentations as well as by word of mouth through elders, opinion leaders, and agricultural extension staff. Suitable dates and venues for public meetings were identified through consultation with local elders and government officials. All people living near the selected study site were invited to the meeting. The agenda of the meeting was described as a discussion of development, health, agricultural, and environmental issues in the area. The date and venue of the public meeting were similarly publicized. Meetings began with self-introductions and an explanation of the objectives. This was followed by an outline of the objectives and methods of the entire project. Participants were asked to share their fears and expectations with regard to the proposed processes and methods and whether they were willing to participate. Dates, time commitment, venues, and other itinerary of initial participatory workshops were discussed and agreement reached.

3.2.3 Initial Village Participatory Workshops

Initial participatory workshops were held in each of the six ISSs with the objective of facilitating residents to describe the study sites systemically in terms of holarchical structure, physical boundaries, resource endowment, institutional structure, historical background, social structure, farming system characteristics, pest and disease dynamics, constraints to human well-being and productivity, and coping strategies. The workshops were held from 7 July to 3 October, 1997. A workshop in each village lasted between 5 and 10 days, depending on the working hours chosen by participants. Facilitators in these workshops were a team of PAR-trained researchers and research assistants from the University of Nairobi and PAR-trained agricultural extension staff and government departmental officials in the district.

After a brief introductory review of the agenda of the workshop, a description of the steps of an action research process and of the objectives and proposed methods of the project was provided. Table 3.1 shows the sequencing of the participatory techniques used in the initial workshops. Transect routes were decided on in a participatory process, with the social and resource maps as a guide. The main criteria for their selection were topography and location of various resources. In all villages,

TABLE 3.1
Sequencing of Learning Tools in the Initial Workshops

Tool	Objectives	Output
1. Self-introduction and icebreakers	Develop rapport Know participants by name Workshop logistics (venue, meals, time)	List of participants by gender Workshop logistics
2. Social and resource mapping	Village boundaries Natural resource inventory Land-use patterns Problem identification	Social map Resource map Lists of identified problems
3. Historical background	Major events and their impacts Problem identification and coping strategies	Historical profile Lists of identified problems Coping strategies
4. Time lines and trend analysis	Resource availability and distribution over time and space Disease and pest dynamics	Graphs of trends and time lines
5. Seasonal calendar	Yearly schedules of activities Yearly trends in climate, agriculture, and pests and diseases	Graphs
6. Transect walks and SSIs	Triangulate resource inventories, problem identification, and social maps Farming system and land use	Graphs of transect profiles showing resource location and land-use characteristics
7. Livelihood analysis	Sources and amount of incomes, types and amount of expenditure	Lists of income and expenditure types
8. Mobility charts	Sources, types, and quantities of goods and services bought or sold	Key inputs and outputs
9. Institutional analysis	Relationships with institutions in the area, their roles and responsibilities Information flow	Venn (chapati) diagrams
10. Daily calendars	Schedule of activities by age and gender Time usage by age and gender Labor distribution by gender	Charts of daily activities by gender and age
11. Health analysis	Health concerns by gender and age Causal structure and coping strategies	Lists of health concerns, their causes, and coping strategies
12. Access and control matrix	Ownership, access, and control of key resources by gender and age	Lists of resources, their ownership, access and control
13. Problem analysis	Types of problems (needs), their causes, and coping strategies	Ranking matrix Causal structure Coping strategies
14. Action planning	Opportunities for remedial action Required inputs, responsibilities, and time frame	Action plans

SSI, semistructured interview

two orthogonal transects were selected. Farm visits and semistructured interviews (SSIs) of farm owners were incorporated into the transect walks. The farms to be visited were purposively selected from a list of households along each transect stratified based on wealth, agricultural practices, natural resource endowment, and ownership. Six to eight farms were selected for each transect route.

In the farm visits, owners or managers were requested to give a guided tour. Special note was taken of the way the owner or manager categorized the various farm enterprises. Farm sketches were made indicating use of the land resource and the types of enterprises. Copies of farm records—if available—were made. A listing of daily time utilization and work schedules of key members (farm owner, spouse, and manager) was made. The owners or managers were asked to explain, in detail, the nature, cause, and severity of existing constraints or problems. For the various farm enterprises, they were asked to give the factors they took into consideration prior to initiating them and what were the essential considerations for continuing those activities. Table 3.2 shows the list of topics covered in the SSIs.

3.2.4.1 Participatory Techniques

The rationale for applying these techniques was to enable communities to describe their situation in details sufficient for the identification and description of problems, issues, and concerns relating to the health and sustainability of their agroecosystem. The primary consideration while selecting techniques for use in this study was that many people in the communities are illiterate to semiliterate, and techniques that involved reading and writing would result in inability (or unwillingness) of the majority to participate in the workshops. The second consideration was that a significant portion of the data came from unwritten formats (e.g., expert or witness statements) and was mostly qualitative. Another consideration was the need for communities to synthesize data into visual representations suitable for viewing and discussion. The techniques used included mapping on the ground or paper; scoring and ranking; interviewing; calendars; Venn diagramming; free listing and card sorting; linkage diagramming; and group presentations and discussions (Chambers, 1994) as well as structured direct observation (Kumar, 1993).

3.2.4.2 Participatory Mapping

Participatory mapping was used to generate spatial representations of various characteristics of the study sites as perceived by the participants and what they perceived to be the boundaries of their community. These provided reference points for data gathering, analyses, and planning in processes similar to those described by Kabutha et al. (1993) and Rietbergen-McCracken and Narayan (1998). The maps were drawn by a group of local participants either on the ground (using chalk, sticks, pebbles, or other available materials) or on large sheets of paper. Two thematic maps were drawn, the first (the resource map) showed the village boundaries and location of various natural resources, while the second, termed the *social map*, showed social factors such as location of various households. Various symbols were used in the

TABLE 3.2
Sequencing of Topics Covered in Semistructured Interviews Conducted in Selected Land-Use Units in Each of the Intensive Study Sites

Topic	Timing of Activity	Expected Outputs
1. Introduction	On arrival	Name of household head Size of household Occupation of household head
2. Land use	Beginning of household/farm/ homestead tour	Settlement history Acreage Ownership, tenure, access, and control of land Apportionment to crops, livestock, dwelling, etc.
3. Crop production and agroforestry	Tour of fields	Types, acreage, and yields by crops and seasons Soil conservation measures Cropping practices (rotation, etc.) Tree types and uses, vegetation
4. Livestock production	Tour of pens and sheds	Production types and yields by species and breed Pest and disease issues
5. Marketing	End of tour of fields, pens, and sheds	Market availability for produce Trends and seasonality of prices
6. Farm inputs	End of tour of fields, pens, and sheds	Types, amount, and costs of inputs (chemicals, labor, seeds, vet services, etc.)
7. Access and control	Beginning of discussion session	Availability, ownership, access, and control of resources Activity profile
8. Institutions	Discussion session	Names and roles or responsibilities of institutions Activities and benefits derived
9. Human health	Discussion session	Common health issues State of health of household members Trends in disease occurrence
10. Livelihood	Discussion session	Sources of income and their relative importance Types and relative importance of expenditures
11. Problems and coping strategies	End of discussion session	Types and relative importance of needs and issues Coping strategies for each

social maps to show household-level characteristics such as relative wealth, levels of resource use, membership in community groupings, and project activity. Discussions of the resource map were geared toward how participants perceived the importance, availability, quality, and utilization of natural resources within the study site.

3.2.4.3 Institutional Mapping

Institutional mapping (Theis and Grady, 1991; Kabutha et al., 1993) was used as a tool to learn about the activities of groups and organizations within the community and to understand how the community viewed these institutions. Local participants generated a list of institutions and individuals perceived to be responsible for decision making in the study site. The perceived relative importance and degree of interaction among the institutions were then depicted in Venn diagrams. First, participants cut out circles from paper to represent each institution or individual. The diameter of the circle indicated perceived relative importance: the larger the circle, the more important the person or institution. A big rectangle was drawn on the ground, on a blackboard, or on paper (depending on the materials chosen by participants) to represent the community, serving as the reference point in the diagram. The rest of the circles were then arranged around this central point with regard to the degree of information sharing and collaboration among them. Separate circles indicated perceived absence of information sharing and collaboration. Touching circles indicated some degree of information sharing between the institutions represented by the circles. Overlapping circles denoted cooperation between institutions, with the extent of overlap indicative of the relative degree of cooperation. Circles inside the rectangle represented those institutions that worked in collaboration with the community. Those outside were seen as important decision makers but without the involvement of the community in their decision-making processes.

3.2.4.4 Historical Background

Historical background was used to outline a brief history and ethnobiography of the people living in the study site. Groups of local participants were divided into groups of 6–10 people, each consisting of at least one representative from different age categories (youth, adults, aged). The oldest member of the group was asked to describe his or her own understanding of where the people in the study site came from and what were the most important highlights in their history. The other participants were asked to add details, seek clarification, or provide alternative viewpoints as the discussion progressed. Each group made a presentation to all participants, and the resulting discussions were recorded.

3.2.4.5 Time Lines

Time lines (Kabutha et al., 1993) provided the community's historical perspective on current issues. Local participants listed historically important events in their chronological order. Time lines were created by groups of 6–10 local participants that included the oldest persons in the study site. The facilitator asked the group to list, in chronological order, the most important events in the history of the people living in the study site. These were followed by group presentations, with general discussions on points of agreement or divergence among the groups.

3.2.4.6 Trend Lines

Trend lines were line plots showing the perceived changes, over time, in key attributes in the study site. In many cases, trend lines were combined with the time lines, the latter forming the horizontal axis of the plot. Groups of local participants, typically 6–10, were asked to show, in a graphical sketch, social, biophysical, and economic changes that they perceived to be the most important in the recent history of the area. Participants were encouraged to graph additional factors deemed important or necessary to explain the trends.

3.2.4.7 Transect Walks

Transect walks (Kabutha et al., 1993; Chambers, 1994) involved walking along predefined routes in the study area and recording differences in soils, land uses, vegetation, crops, livestock, and use of technologies. The aim was to visually appraise the status of the village and its resources to better identify and assess problems, solutions, and opportunities. Findings were recorded in a representational diagram, showing a cross section of the study site along the transect route and the extent of ecological, cultural, or economic subzones within the study site. Differences between zones in terms of problems and opportunities were also highlighted in the diagram. Transects were carried out by a team of local (about four) and external participants (usually two). In this study, they combined semistructured and unstructured interviews with residents and farmers along the route. Two to four routes were selected (depending on the size of the study site and zoning pattern) based on the main geophysical and social factors identified in the mapping exercise.

3.2.4.8 Semistructured Interviews

The objectives of SSIs (Chambers, 1994; Rietbergen-McCracken and Narayan, 1998) were (1) to learn about a particular situation or group in detail, (2) to discuss issues that would have been difficult to address using other methods, and (3) to reveal personal perspectives on particular topics. SSIs, also called conversational interviews, were used in several contexts in this study. The first was in the description of villages and their problems, coping strategies, and opportunities. These SSIs were carried out together with the transect walks. Interviewees in this case were individual community members and farmers selected through a stratified sampling process based on wealth ranking and household characteristics such as size and gender of household head, supplied by the participants in the mapping exercise.

In other applications, interviewees were special interest groups or key informants, depending on the purpose of the interview. In all cases, interviewers were provided with a checklist of topics as a guideline. Interviewers were asked to remain conversational enough to allow participants to introduce and discuss issues that they deemed relevant. In some cases, visual-aid-based methods were used as opposed to the more traditional verbal methods. Visual aids were used more often in group interviews and in the application of soft system methodology. Interviews were conducted by a team of two to four people in an informal setting that allowed mixing of questions and discussion while avoiding leading questions, questions with yes-no answers, and value judgments. These interviews were restricted to 45 minutes or less (Theis and Grady, 1991).

3.2.4.9 Seasonal Calendars

Seasonal calendars (Theis and Grady, 1991; Kabutha et al., 1993; Rietbergen-McCracken and Narayan, 1998) were diagrams showing perceived annual trends in various biophysical and socioeconomic phenomena in the study sites. Seasonal calendars were often drawn on the ground, with relative trends depicted using stones and seeds, but in some cases pen and paper were used to draw simple line graphs showing seasonal increases and decreases. Several variables, such as pests and diseases, crop yields, and labor, were included in the calendar to enable an assessment of relative annual patterns. Seasonal calendars were drawn by local participants assembled in groups structured to include different ages, gender, and leadership perspectives as described by Kabutha et al. (1993). The facilitator asked participants to mark out the year into seasons using their local language and to use preferred media to mark out trends in selected biophysical and socioeconomic variables.

3.2.4.10 Daily Activity Charts

Daily activity charts (Chambers, 1994; Rietbergen-McCracken and Narayan, 1998) were created to show daily time use for the average individual in the community and to show the types of routine activities, relative amounts of time spent on them, and degrees of drudgery. Daily activity charts were made by focus groups categorized by gender, age, employment, and marital status. Group presentations were done to elicit intergroup perspectives. Comparisons of the daily activities of different groups were made and discussed.

3.2.4.11 Focus Group Discussions

Focus group discussions (D'Arcy, 1990; Kumar, 1993; Cabanero-Verzosa et al., 1993; Rietbergen-McCracken and Narayan, 1998) were used as a means of obtaining in-depth information on a specific topic through a discussion. Focus group discussions were designed as facilitated discussions on a specific topic by a small group of people who shared common concerns. Participants discussed ideas, issues, insights, and experiences among themselves, and each member was free to comment, criticize, or elaborate on the views expressed by others. It was not expected that participants would have only one opinion or that they would agree on anything, but rather that the similarity of their orientation toward the issue at hand would allow free sharing of information and deeper insight into the issue under discussion. The goal of the facilitator was to create a situation in which the participants were stimulated to talk with each other on the chosen topic. The primary role of the facilitator was to stimulate group discussion, to keep discussions within reasonable limits of the topic at hand, and to prevent a few participants from dominating the discussions. Focus groups were limited to between 8 and 12 participants. The small size of the group was intended to facilitate the free flow of discussions. A session generally lasted between 1 and 2 hours. Several sessions with different participants were held on a specific topic.

3.2.4.12 Presentations and Analysis

In group presentations, participants in group activities such as mapping or transect walks made a presentation on their findings to the rest of the workshop participants. The objectives were to review the outputs of the group activity for accuracy and completeness, to analyze the data generated, and to stimulate expression of convergence or divergence perspectives on issues brought out by the group activity. Group presentations were held at the end of a group activity. Participants were requested to review the outputs of their group, prepare visual aids, and decide on a mode of presentation. Several members of the group were selected to present various topics or aspects of the outputs. The group presentation forum was similar to a public lecture, with questions and comments reserved until the end of the presentation, followed by a general session at which comments were made by other workshop participants.

3.2.4.13 Wealth Ranking

Wealth ranking (Grandin, 1988; Chambers, 1994; Rietbergen-McCracken and Narayan, 1998) was used to rank households according to their perceived well-being or wealth. The objective was to reveal potential socioeconomic stratifications of the population and to identify local people's definitions and criteria of wealth and well-being. This technique involved a series of individuals or focus groups of local participants ranking the entire community based on predefined criteria. Facilitators introduced the technique using local terms for wealth and poverty, encouraging participants first to discuss how they define these terms. Subsequently, local participants were asked to list the criteria they would use to classify a household or individual as poor or rich. If many, divergent criteria were given, pairwise ranking was used to determine the most important of these. If possible (based mainly on time constraint), ranking was repeated serially with different people and the results compared, looking for any large discrepancies or differences in the classification of the households, especially in the proportion of households in each of the categories. The actual ranking was done using card sorting. First, participants constructed a list of all households to be ranked. The name of each household was written on a separate card. The person or group doing the ranking was asked to sort the cards into three groups (poor, average, and rich) based on his or her perception of each household's wealth and well-being status and using the predefined criteria. The actual proportions of households in each category were recorded for each ranking exercise and then averaged.

3.2.4.14 Health Analysis

Health analysis begun by a listing of health issues deemed to be the most important in the village. Local participants were assembled into age- and gender-specific groups for this. The lists were then compiled onto sorting cards and a pairwise ranking carried out to identify the most important of these. Gender differences, if any, were noted and discussed in a group presentation forum. For each of the most important health issues identified, the causes, coping strategies, and opportunities were identified.

3.2.4.15 Problem Identification and Ranking

Problem ranking was used to assess the relative importance of problems, issues, and concerns as perceived by the local participants. An initial list of problems, issues, and concerns in the study site were constructed through triangulation. Triangulation involved facilitators reexamining all the outputs (maps, charts, and tables) from the workshops and listing themes, issues, and concerns identified as problems or constraints. The relative frequency of a particular theme, issue, or concern was seen as an indicator of its relative importance. Problems and concerns mentioned in only one of the outputs were not included in the initial list. Local participants were then asked to add any other problem or concern that they thought should be included. After participants confirmed that the list was exhaustive, the problems were listed on sorting cards and a pairwise ranking carried out. In the pairwise ranking, the facilitator showed the cards two at a time, each time asking the participants to decide which of the two concerns depicted was the bigger problem to the residents. A tally mark was made on the back of a card whenever the concern it depicted was chosen. The cards were then sorted in order of the tally marks, the lowest card having the fewest tally marks and the topmost card having the most.

3.2.4.16 Problem Analysis

In this process, the perceived causes, the coping strategies, and the opportunities for resolution of stated problems or concerns were assessed. A tabular matrix was drawn on the ground using chalk or on a large sheet of paper using felt pens. The first column identified the problems or concerns. The subsequent columns identified the analytical themes (causes, coping strategies, opportunities). Each problem row in the table represented a problem, ranked in the order of severity as identified in the pairwise ranking. Each of the most important problems was analyzed from each thematic viewpoint and the outputs recorded either pictorially or using descriptive statements in the tabular matrix. Problem analysis was carried out in groups of 6–10 local participants. Group composition in terms of gender, age, and other criteria depended on the nature of problems analyzed.

3.2.4.17 Preference Ranking

Preference (Rietbergen-McCracken and Narayan, 1998) ranking involved the assessment of options based on predefined criteria. It was carried out using card sorting similar to that in problem ranking. In this case, the facilitator asked participants to identify the better of two options. Preference ranking was used to identify the most suitable opportunities for each of the problems.

3.2.4.18 Action Planning

Action plans were activities—listed in order of priority—that were to be undertaken to meet defined goals and objectives. Also included was a list of resources needed to complete the tasks, sources of funds and materials, and the actors for each activity listed.

3.2.5 Follow-up

Follow-up workshops were scheduled every 3 months to monitor the implementation of action plans and annually to carry out evaluations, replan research and development activities, and carry out agroecosystem health assessments. The choice and sequencing of participatory tools varied depending on the objectives of the workshop.

3.2.5.1 Creating Cognitive Maps

Cognitive maps (also known as loop models, influence or spaghetti diagrams) are models that portray ideas, beliefs, and attitudes and their relationship to one another in a form that is amenable to study and analysis (Eden et al., 1983; Puccia and Levins, 1985; Ridgley and Lumpkin, 2000). Cognitive maps were developed, one for each intensive study site, in 1-day participatory workshops. Participants were divided into groups of 6–10, and each group was requested to show how various social, economic, and biophysical factors influenced the health and sustainability of their agroecosystem. Group activities were followed by group presentations in which intergroup discussions were recorded. Details of the methods and processes used are described in Chapter 4.

3.2.5.2 Developing Community-Based Indicators

Community-driven indicators were developed through a participatory process in which communities in six study sites were asked to list things that they would measure to determine if their agroecosystem was becoming more or less healthy or sustainable. Details on the selection of study sites and the participatory tools used in this process are provided in Chapters 2 and 6, respectively.

3.2.5.3 Monitoring, Evaluation, Planning, and Assessments

For those indicators considered suitable, the tools, methods, resources, and time frame needed for carrying out measurement were debated and agreed on. Four to six groups of participants were formed to carry out measurement of different indicators grouped on the health attributes for which they provided most information.

In each of the six villages, measurements were carried out over a period of 3–4 weeks. During this period, groups charged with measurement of specific indicators within each village met weekly to discuss progress and results. After all groups in a village had completed the measurement process, a 1-day workshop was then held in the villages, and each of the groups presented their findings. Participants were encouraged to debate the state of health (whether poor, average, or good) of their agroecosystem and to state the reasons why. They were also asked to debate whether the health was improving, deteriorating, or steady. Subsequently, communities preferred to carry out the assessments during intervillage meetings.

3.3 RESULTS

Community participation in PAR workshops was high, with 75% to 100% of the households and homesteads represented in all the participatory workshops held in the study sites. In all the communities, the concept of participation in a research process was new, but the concepts underlying the research were reported to be similar to traditional practices used by farmers and artisans. The use of tools that removed the need for literacy was considered useful by the communities. Two cultural factors, however, influenced the quality and detail of data on some topics. The most affected were causes and degree of mortality and wealth ranking. In all the communities, the participants conceded that they were unable to discuss in detail issues related to mortality due to cultural values that prohibited discussions of mortality in public. Participants were reluctant to talk about wealth (common and individual) as this was tantamount to "telling God that you have had too much to eat." The concept of agroecosystem health was well understood by most community members, as evidenced by their use of health language, images, and concepts throughout the participatory workshops.

3.3.1 Community Identities

3.3.1.1 Participatory Mapping

Based on the descriptions by participants, individuals in communities in the ISSs perceived themselves as residents of a village with well-defined boundaries and membership. In the Kiawamagira ISS, the village was described as the area along the Nairobi-Kikuyu road and bounded on the southern and western sides by the Nairobi-Kisumu railway line. The village was described as consisting of 60 households and homesteads. For the purpose of this study, boundaries of the study site were changed to correspond with those described by the participants.

3.3.1.2 Institutional Mapping

Table 3.3 shows the institutions considered by communities in the ISSs as important in decision making and the relationships among them and with the communities. All communities indicated that administrative officials were important in decision making, but two study sites (Githima and Gitangu) indicated lack of a collaborative relationship between the community and the administration. Only one village (Gitangu) indicated that there were relationships between institutions at the community level. Other communities indicated that these institutions operated independently. All villages except Gikabu-na-Buti indicated a collaborative link between the community and schools.

3.3.1.3 Historical Background and Time Lines

Table 3.4 presents a summary of the historical backgrounds and time lines given by participants in the six ISSs. Gikabu and Thiririka were recent settlements, the former consisting of tea estate workers who bought a portion of land from a tea estate and subdivided it among themselves. Thiririka, the youngest of the villages, is a settlement of former squatters in government forests. Although the community in Gitangu village is divided into three different clans, they have a history of working together as a unit. In Gikabu, the two groups that existed did not work together at all despite presenting themselves to outsiders as a unit. Information was restricted, and most projects were managed by each group separately. Gikabu and Thiririka had the shortest time lines as the most recently settled. All time lines revealed a concern with biophysical phenomena, especially related to food production.

TABLE 3.3
Institutions Perceived to be the Most Important in Decision Making Among the Intensive Study Sites, Kiambu District, Kenya, 1997

Institution	Githima	Gitangu	Kiawamagira	Mahindi	Gikabu	Thiririka
Administration[a]	D	D	C	C	C	C
Agriculture[a]	D	I[1,2]	D	D		I
Churches	C	I[5]	C	C		I
Coffee factories	C[1]					
Co-op	C[1]	I[1]			I	
Culture[a]		I[3,4]				
Plan International		I[2,3]			C	
Police	D					
Politicians		I[5]				
Private clinics	I					
Private vets	I		C	C	C	
Health[a]	D		D	D	I	C
Public works[a]		I[1]	D	D		
Schools	C[2]	C[2,3,4,5]	C	C	I	C
Tea centers	C[1]					
CBOs	C[2]	C[3,4,5]				D
Forestry[a]			D			I
Health foundation					C	
World vision						D

[1,2,3,4,5] C, collaboration between the institution and village community; D, decision making only; I, information and some degree of interaction with the community

[1,2,3,4,5] Institutions with similar superscript numbers perceived as collaborating with each other by residents of the respective villages

[a] Government departments

3.3.2 Profiles and Trends

3.3.2.1 Trend Lines

Attributes included in trend lines are summarized in Table 3.5. In Githima village, rainfall and soil fertility were perceived as decreasing since 1964, resulting in declining crop and livestock yields. Both phenomena were seen to be related to the cutting down of the forest that once existed in the area. The number of people engaged in farming as well as the intensity of farming were reported to have been increasing since the early 1960s. Scarcity of farmland became an issue beginning from the early 1970s, and this was seen as resulting from an increasing population growth rate since the late 1950s. The increase in human diseases, of which pneumonia and colds were the most common, was associated with lack of water, an increase in the use of agricultural chemicals, smoking, and a changing lifestyle.

TABLE 3.4
Historical Backgrounds and Time Lines as Presented by Residents in Six Study Sites in Kiambu District, Kenya

	Githima	Gitangu	Mahindi	Thiririka	Kiawamagira	Gikabu
Meaning of name	"A natural water spring"; there existed such a spring before settlement	No ascribed meaning; the village derives the name from a water spring located in the area and the river that flows from it	"Bones"; the village is littered with elephant bones; reputed to have been an elephant graveyard	No ascribed meaning; derived from the name of the stream flowing through the village	"Wamagira's village"; village reputed to have been the hideout of a cattle rustler named Wamagira prior to settlement	"Basket and measuring rod"; residents are mostly tea pickers; the basket and the measuring rod are the "tools of trade" of a tea picker
Settlement	Prior to 1900	Prior to 1900	Early 1950s	1989	Late 1950s	1972
Origin of inhabitants	Surrounding villages	Murang'a	Nearby village called Kihara	Squatters in Kamae, Kieni, and Kinale forests	Squatters in the Church Missionary Society lands in Thogoto	Tea pickers in surrounding tea estates
Initial Status	Wattle tree forest	Forest	Forest	Forest	Swamp	Tea estate
Community groupings	None	Three clans (mbari ya Igi, Mbari ya Ngoru, and Mbari ya Gichamu)	None	Three groups based on the forest in which they were squatters before settlement (kamae, kinale, kieni)	Two (outsiders vs. insiders); outsiders are those who bought land from original inhabitants	Two based on farm size (Gikabu and Itungi); Itungi has 5-acre plots, while Gikabu has half-acre plots
Relationship among groups		Excellent		Poor; many activities organized at group level	Moderate; outsiders said to be reluctant to participate in village activities	Poor; these are separate communities

(continued on next page)

TABLE 3.4 (continued)
Historical Backgrounds and Time Lines as Presented by Residents in Six Study Sites in Kiambu District, Kenya

	Githima	Gitangu	Mahindi	Thiririka	Kiawamagira	Gikabu
Time lines	1948 Locust invasion; clearing of forest 1952 Emergency state 1954 Tea and coffee planting 1961–1997 Floods, decreasing rainfall, population growth, improved hygiene, new diseases, high incidence, better health care	1919 Kimiri (smallpox), forced labor 1941 Plague 1943 Famine 1949 Foot-and-mouth disease; Exotic trees and crops 1952 Emergency state, forced into villages 1957 Land demarcation 1961 Floods 1964 Famine, yellow maize 1970 New ticks species 1973 New varieties of weed 1984 Famine, drought, yellow maize, first AIDS case	1928 Karura church built; hospital built at Kikuyu town 1943 Famine 1950 Mau Mau uprising, accelerated clearing of forest 1958 Land demarcation 1959 Permitted to grow coffee, increasing farm productivity, decreasing farm sizes	1988 Land allocation to squatters 1989 Settlement into village 1990 Bumper harvest 1992 Famine, drought, councilor elected, nursery school built 1994 Primary school started; severe frost, most crops destroyed, famine 1996 Drought, food shortage	1920 Missionaries settle at Thogoto; people became squatters 1936 Flooding 1948 Railway construction 1952 Forced relocation 1958 Land demarcation 1960 Population growth 1970 Grade cattle introduced 1973 Famine 1975 Nairobi-Kikuyu road 1978 Gikambura road constructed 1980 Water project started 1984 Famine, drought, yellow maize 1989 Heavy flooding, derailment 1990 Electricity installed	1960 White settlers move out 1961 Flooding 1963 Independence 1964 Famine, yellow maize 1970 Title deeds issued 1972 Settlement into village 1976 Nursery school started 1978 Death of Jomo 1980 Nazareth-Limuru road 1982 Bus services 1984 Famine, drought 1986 Tea first grown 1989 Plan International 1996 Tigoni becomes division 1997 Famine, relief food

TABLE 3.5
Attributes Included in Trend Line Diagrams, Kiambu District, Kenya, 1997

Attribute	Githima	Gitangu	Kiawamagira	Mahindi	Gikabu	Thiririka
Availability of farmland	D		D	D		
Availability of firewood		D	D	D		
Crop diseases		I	I			I
Crop yield	D	I	D		D	
Education				I		
Emigration				I		
Extension services	D				D	
Farm sizes	D		D	D		
Farming intensity	I				I	
Flooding			I		I	I
Food production			D	D	D	
Human diseases	I		I			I
Human population	I		I	I	I	
Land under cultivation	I		I		I	I
Livestock diseases						I
Livestock numbers					I	
Livestock yield	D	I			D	
Number of farmers	I					
Rainfall	D	D				D
Soil erosion	I				I	I
Soil fertility	D	D	D	D	D	
Traditional crops		D				
Traditional livestock breeds	D	D				
Tree cover	D				D	D
Unemployment				I		
Use of agrochemicals	I				I	
Water for domestic use	D		D			
Water quality			D			

Blank, not included; D, decrease; I, increase

In Gitangu village, the most significant trends were reported to be a decline in soil fertility, change in the types of crops and livestock produced, reduction in rainfall, and a decline in the availability of firewood. Sorghum, sweet potatoes, millet, njahi (*Dolichos lablab*), bananas, cassava, maize, and beans were reported to have been the major crops in the 1930s. Production of these crops declined from the 1940s, reportedly due to changes in dietary preferences. Sorghum, millet, and njahi were no longer produced. Pyrethrum was introduced in 1947, but production ceased in the early 1970s due to low prices. Sweet potato production was severely hampered by pests (moles and termites) as was production of beans (weevils) and Irish potatoes (blight and bacterial wilt). The introduction of hybrid maize seed was reported to be the major factor in the perceived increase in maize production. The declaration of a state of emergency in 1952 was said to be the main cause of the changes in livestock production because of restriction in grazing activities, thefts, heavy taxation, and fines. Sheep and goat production dropped and remained at a low level. In contrast, cattle and poultry production had been increasing. Donkeys had been reintroduced more recently for the purposes of transporting goods and water.

Residents of Kiawamagira village reported an increase in population, increase in the land under cultivation, and a decline in water availability as the most important trends since the 1950s. The consequences of these were a decline in soil fertility, declining crop yields, increasing human diseases, small farm sizes, and lack of firewood. Rising population, increased level of education, and decreased farm sizes were reported by Mahindi residents. Important trends in Gikabu were declining soil fertility, declining tree cover, increasing population, low crop yields, rising population, increased land under cultivation, increased soil erosion, and poor extension services since the late 1970s. Beginning from 1990, there was reported to have been a decline in rainfall, accelerated clearing of forest and woodlots, increased flooding and soil erosion, increase in land under cultivation, and increase in human, crop, and livestock diseases by residents of Thiririka village.

3.3.2.2 Transect Walks and Semistructured Interviews

Table 3.6 shows results of transect walks and SSIs conducted during the initial workshops. Maize, kale, sheep, and dairy cattle production were observed in all six study sites. Mahindi had the least reported cash-earning produce (two) followed by Gikabu-na-Buti (four), while Githima had the most (six). Stream water was used for domestic purposes in all the villages. Farmers in all the villages who were visited during transect walks reported using commercial fertilizers (chemical).

3.3.2.3 Wealth and Well-Being

The criteria used by local participants to categorize households based on their wealth and well-being status are shown in Table 3.7. Type of house and size of farm were used by participants in all study sites. In Githima, a wealthy household was described as one in which members owned houses, cars, shops, or shares in companies, were able to afford education for their children, had high personal hygiene, and had a well-managed farm. Participants reported that 10% of the population in Githima can be said to be wealthy, 50% average, and 40% poor.

A household that lived in a permanent house and with members who owned vehicles or ran businesses was described as the wealthiest in Gitangu. Only 10 households in Gitangu could be described as wealthy. Local participants estimated that 5% of households in Gitangu were poor.

Households with small farms located on marshy, hilly, or stony areas; unable to purchase farm inputs; living in nonpermanent houses with a shortage of water; and with none of the members having off-farm employment were described as poor in Kiawamagira. Of the households in Kiawamagira, 45% were described as poor, while the rest were described as "not poor."

TABLE 3.6
Results of Transect Walks and Semistructured Interviews, Kiambu District, Kenya, 1997

Variable	Githima	Gitangu	Kiawamagira	Mahindi	Gikabu	Thiririka
Types of Crops						
Tea	XO				XO	
Coffee	XO					
Maize	XO	XO	XO	XO	XO	XO
Beans	XO	O	XO	XO	XO	
Potatoes	XO	O		XO	XO	XO
Kales	XO	XO	XO	XO	XO	XO
Tomatoes	XO	XO	XO			
Bananas	O	O	O	O	O	
Napier		O	O	O	O	
Arrowroots			O			
Sugarcane				O		
Flowers				XO		
Peas					XO	XO
Carrots						O
Crop Pests and Diseases						
Coffee berry disease	X					
Stock borer	XO		XO		XO	
Bacterial wilt	XO	XO	XO		XO	
Cutworm		XO	XO		XO	XO
Blight			XO		XO	XO
Voles			XO		XO	
Maize streak					XO	
Spider mites						XO
Soil Conservation Measures						
Terraces		O			O	
Gabions		O	O			
Grass strips		O	O		O	
Cropping Practices						
Intercropping	O	O	O	O	O	
Crop rotation						XO
Tree Types						
Fruit trees	O	O	O	O	O	
Woodlots	O			O	O	
Agroforestry					XO	XO
Livestock						
Cattle	O	O	O	O	O	O
Sheep	O	O	O	O	O	O
Goats	O			O		
Bees	O					

(continued on next page)

TABLE 3.6 (continued)
Results of Transect Walks and Semistructured Interviews, Kiambu District, Kenya, 1997

Variable	Githima	Gitangu	Kiawamagira	Mahindi	Gikabu	Thiririka
Range chicken	O		O	O		
Commercial poultry		O				
Livestock Pests and Diseases						
Foot-and-mouth disease	X					
Rinderpest	X					
Ndigana[a]		X				X
Infectious bursal disease of chicken		XO				
Mastitis				X	X	X
Ticks				X		
Pneumonia					X	X
Foot rot					X	X
Worms					X	
Cash-Earning Produce						
Coffee	X					
Tea	X				X	
Milk	X	X	X	X	X	
Kales	X	X	X		X	X
Tomato	X		X			
Maize	X		X			
Eggs		X				
Poultry		X				
Flowers				X		
Carrots						X
Peas						X
Potatoes						X
External Farm Inputs						
Labor	O	X			X	
Fertilizer	X	X	X	X	X	X
Manure	X		XO	X		
Fungicides	X					
Pesticides	X		X			XO
Seeds			X	X		X
Fodder	O	XO			XO	
Sources of Water for Domestic Use						
Boreholes	O	O				
Rainwater	O	O	O	O	O	
Shallow wells		O		O	O	O
River	O	O	O	O	O	O
Tap			O			

TABLE 3.6 (continued)
Results of Transect Walks and Semistructured Interviews, Kiambu District, Kenya, 1997

Variable	Githima	Gitangu	Kiawamagira	Mahindi	Gikabu	Thiririka
		Problems, Needs, Issues, or Concerns				
Market[a]		X		X	X	
Water		X				
Crop diseases			X			
Animal diseases[b]		X				
Artificial insemination services			X			
Soil fertility			X			
Soil erosion	O		XO	O	XO	

O, observed; X, reported

[a] Market unavailable or unstable prices for produce

[b] Often used in reference to anaplasmosis, but also to other conditions presenting with constipation or indigestion

TABLE 3.7
Criteria for Judging Wealth and Well-Being of Households, Kiambu District, Kenya, 1997

Criteria	Githima	Gitangu	Kiawamagira	Mahindi	Gikabu	Thiririka
Type of house lived in	X	X	X	X	X	X
Motor vehicle ownership	X	X		X	X	
Proprietor of a business	X	X			X	
Off-farm jobs		X	X	X	X	
Size of farmland	X	X	X	X	X	X
Owns shares in companies	X					
Educates/d children	X			X	X	
Personal hygiene	X			X		
Farm management	X		X			
Quality of diet	X		X	X		
Health status of family	X					
Electricity supply to home				X		
Telephone service at home				X		
Livestock numbers						X
Bicycle ownership						X
Television ownership						X

In Mahindi, wealthy households had permanent houses, cars, a telephone, and electricity; more than 3 acres of farmland; educated children; and a well-fed family. Wealthy households in Thiririka were those that owned land and a bicycle or television set; had at least two cows, a good timber house, and a wife and children. Approximately 20% of the residents in Thiririka were described as poor, 75% as average, and 5% as wealthy. Local participants in Mahindi and Gikabu could not give estimates of the number of rich and poor households, citing sociocultural reasons.

3.3.2.4 Health Analysis

Diseases and health concerns described by local participants as the most important and commonly occurring are shown in Table 3.8. The most important risk factors described by local participants are shown in Table 3.9; the coping strategies for these problems are shown in Table 3.10.

TABLE 3.8
Diseases and Health Concerns Perceived to be the Most Important and Commonly Occurring, Kiambu District, Kenya, 1997

Diseases	Githima	Gitangu	Kiawamagira	Mahindi	Gikabu	Thiririka
Malaria	X	X	X	X		X
Dysentery	X		X			
Pneumonia	X	X	X	X	X	X
Coughing	X					
Typhoid	X	X	X		X	
Flu/common cold	X	X	X	X	X	X
Asthma	X	X				X
Backache	X					
High blood pressure	X	X				X
Stomach ulcers	X					X
Diabetes	X	X				X
Tuberculosis		X				
Joint pains/ arthritis		X		X	X	X
Cancer		X				
AIDS		X	X		X	
Skin diseases/ ringworms		X			X	X
Epilepsy		X				
Diarrhea					X	
Gastrointestinal tract worms						X

TABLE 3.9
Factors Perceived as Increasing the Risk of Diseases, Kiambu District, Kenya, 1997

Risk Factors	Githima	Gitangu	Kiawamagira	Mahindi	Gikabu	Thiririka
Mosquitoes	Malaria	Malaria		Malaria		
Cold weather	Flu Colds/flu	Malaria Colds/flu	Pneumonia Colds/flu		Pneumonia	
Dust and pollen	Asthma	Asthma Colds/flu				
Strenuous work	Backache Arthritis Joint pain					
Stress	HBP Ulcers Diabetes					HBP Ulcers
Dietary change	Ulcers					
Swamps and flooding		Malaria Colds	Malaria			
Genetic susceptibility		Diabetes Asthma				
Poor hygiene		Tuberculosis				
Drinking polluted water		Typhoid	Typhoid			
Bad morals			AIDS			
Inadequate nutrition			All[a]			
Agrochemicals	All[a]					All[a]
Old age					Joint pains Arthritis	

HBP, high blood pressure
[a] Factor considered to increase susceptibility to all kinds of diseases

3.3.3 Problem Analysis and Action Planning

3.3.3.1 Problem Identification, Ranking, and Analyses

A summary of problems and concern as ranked by participants in the initial village workshops is given in Table 3.11. Concerns common to all the villages were availability of water for domestic use, poor roads, and poor health and health facilities. Only one village (Kiawamagira) had access to piped water, available for one half-day per week. Roads were mainly loose surface, becoming impassable during the wet season. Due to the hilly terrain of the agroecosystem, flooding and soil erosion were the biggest causes of poor road conditions. Among the agriculture-related problems

TABLE 3.10
Strategies Used to Cope with Human Diseases, Kiambu District, Kenya, 1997

Strategy	Githima	Gitangu	Kiawamagira	Mahindi	Gikabu	Thiririka
Consulting a doctor	X	X	X	X	X	X
Consulting a herbalist	X					
Using herbs/ traditional remedies	X	X			X	X
Using over-the-counter medicine					X	
Avoiding drinking polluted water		X				
Improving hygiene		X				
Boiling water before drinking			X			
Vaccination					X	
Improved nutrition			X			

were lack of artificial insemination services associated with poor state of roads, low crop yields, poor soil productivity, lack of markets, lack of extension services, and crop and animal diseases.

3.3.3.2 Preference Ranking and Action Planning

Based on their initial agroecosystem health diagnosis, communities developed action plans and the organizational structures to carry these out. The action plans developed by the six ISSs are summarized in Table 3.12.

3.3.4 Follow-Up

3.3.4.1 Collective Action

At the end of the initial village workshops, all communities expressed a profound demand for action to ameliorate the problems identified. Formation of the village committees was evidence of their desire to implement the action plans. Five of the six villages proceeded with implementation of the action plans immediately after the workshops, mostly without further contact or consultation with the research team. In nearly all the cases, this led to some degree of frustration on the part of the communities as they were ill-prepared in terms of organization and community leadership to carry out many of the tasks. However, there were some successes and failures, and frustration did not deter most of the communities from continuing to try.

3.3.4.2 Reflection and Replanning

Table 3.13 shows the list of revised action plans and the progress in their implementation from 1997 to 2000. Githima village revised its action plans to begin with road

TABLE 3.11
Summary of Problems and Concerns as Prioritized by Participants of the Initial Village Workshops in the Intensive Survey Site, Kiambu District, Kenya, 1997

Rank	Githima	Gitangu	Kiawamagira	Mahindi	Gikabu	Thiririka
1	Water not easily accessible	Water not easily accessible	Poor roads	Poor roads	Water not easily accessible	Security inadequate
2	Poor human health and health care	Poor roads	Water shortage	Unemployment	Security inadequate	Poor human health and health care
3	Illiteracy	Poor human health	Low farm productivity	Water not easily accessible	Poor health and health care	"Grabbing" of public land
4	Poor roads	Unemployment and crime	Fuel shortages	Poor human health and health care	Unemployment	Poor-quality seeds
5	Fuel shortages	Secondary school and polytechnic needed	Security inadequate	Nursery school needed	Crop diseases	Lack of unity and solidarity
6	Lack of artificial insemination services	Crop diseases, pests, and poor seed quality	Inadequate AI services	Lack of knowledge	Outlet for tea produce needed	Lack of extension services
7	Security inadequate	Animal diseases and poor-quality feeds	Poor human health and health care	Livestock disease	Lack of extension services	Poor leadership
8	"Ignorance"	Soil erosion and infertility	Poor communication		Alcoholism and drug abuse	Improper use of agrochemical
9		Lack of market and shopping center			Lack of school fees	Soil erosion
10		Inadequate extension services			Food shortages	Crop diseases

TABLE 3.12
Actions Planned by Communities in Intensive Survey Site, Kiambu District, Kenya, 1997

Githima	Gitangu	Kiawamagira	Mahindi	Gikabu	Thiririka
Start a self-help medical clinic	Rehabilitate Gitangu water project	Carry out road repairs and regular maintenance	Carry out road repairs and regular maintenance	Construct a village dam	Organize security meetings
Carry out road repairs	Carry out regular maintenance	Start a water supply project	Start income-generating activities	Organize security meetings	Organize health training, rehabilitate health center
Rehabilitate water system	Start a mobile medical clinic	Request for extension services	Start a water supply committee	Improve existing dispensary	Request for extension services
Add classrooms to secondary school	Organize community security groups	Promote energy savers and agroforestry	Start a village medical clinic	Start self-help projects	Seek title deeds for public utility lands
	Start a village polytechnic	Start village security groups	Start a village nursery school	Seek extension services	Start a water supply scheme
	Start village extension programs	Start a community dispensary			Form small marketing cooperatives

rehabilitation, electrification, water supply, expansion of school, and then development of an extension program.

Gitangu village opted to leave water supply as the first priority, but extension was moved to the second place. Soil conservation was given first priority in Kiawamagira, followed by road rehabilitation and control of flooding. Mahindi retained only two items in their action plans: (1) rehabilitation of the access road and (2) development of an income-generating project. Gikabu-na-buti village revised its action plans to electrification, water supply, income generation, and extension. Thiririka reordered its action plans to begin with water supply, followed by development of a market for farm produce, extension, and then road rehabilitation.

3.4 DISCUSSION

The combination of the health language and participatory methods provided a means for the communities to make qualitative evaluations of their agroecosystems. There were, however, important constraints to this process, mainly stemming from cultural practices that inhibited public discussion of many of the issues impacting negatively on the health and sustainability of these agroecosystems.

TABLE 3.13
Revised Action Plans and Progress in Implementation, Kiambu District, Kenya, 1997–2000

Village	Githima	Gitangu	Kiawamagira	Mahindi	Gikabu	Thiririka
Planned activities in the revised action plans	Road, electricity, water, school, extension	Water, extension, security, medical	Conserve soil, road, control flood, extension, security	Road, income generation	Electricity, water, security, income generation, extension	Water, market, extension, road
Number of projects past initiation stage by October 1998	All	Security, extension	Security, soil conservation	Income generation	None	Road, extension
Number of activities completed by October 1998	Electricity, water	Security, extension	Security, soil conservation	None	None	Road
Number of projects past initiation stage by August 2000	Expansion of school	Install water meters	Road and flood control	None	None	None
Activities current as of August 2000	School expansion	Water meters	Road and flood control	None	Electricity supply	Water supply
Stage of project considered current as of August 2000	Near completion	Beginning	Beginning	Not started	Planning	Planning
Number of planned activities completed by August 2000	All	All	Extension, soil conservation	Income generation	Extension workshops	Road
Recurrent activities as of August 2000	Road maintenance	Extension meetings, M&E meetings	Road maintenance, soil conservation meetings	Road maintenance	Nil	Nil
Frequency of participatory meetings	Frequent[a]	Very frequent[b]	Rare[c]	Rare	None	Rare
Attendance at meetings	High[d]	High	Moderate[e]	Poor[f]	Poor	Moderate

(continued on next page)

TABLE 3.13 (continued)

Revised Action Plans and Progress in Implementation, Kiambu District, Kenya, 1997–2000

Village	Githima	Gitangu	Kiawamagira	Mahindi	Gikabu	Thiririka
Linkage with other intensive survey sites	Very high[g]	Very high	Moderate[h]	Moderate	High[i]	None
Funds generated (by community) to support activities (in Kenya shillings)	1.2 million	120,000	10,000	6,000	Nil	20,000

M&E = Monitoring and Evaluation

[a] Roughly once every 2 months.

[b] Twice a month.

[c] Not regular; only when need arises.

[d] More than half of the households represented.

[e] More than a quarter but less than a half of the households represented.

[f] No more than a quarter of the households represented.

[g] Have initiated visits to other villages and to their own village.

[h] Have initiated and hosted visits to their village.

[i] Have participated in all intervillage meetings.

3.4.1 Community Identities and Collective Action

All communities showed a strong sense of identity, with little or no sharing of resources across village boundaries. Resources across the boundaries shown in their resource maps were described as belonging to other people. Associations, burial committees, women's groups, and other community institutions were often contained within the village boundaries. The existence of villages as a level of organization in the human activity holarchy was further confirmed by the apparent integrated sharing of resources and reciprocal exchange of means of production and information in addition to the apparent shared sense of belonging together. Although the interaction between researchers and the communities in extensive villages was minimal, similar organizational characteristics were suspected to exist based on descriptions by key informants.

In spite of the existence of these communities, the ability to implement action plans (Table 3.13) differed significantly among the villages, indicating that this was a result of factors other than the sense of belonging. An interesting feature was that the villages with a high degree of cooperation (based on the level of participation in village activities; Table 3.13) described their relationship with the administrative arm of government as that of decision makers rather than collaborators (Table 3.3). Furthermore, these villages were reported as having more nongovernmental institutions that collaborated with each other (Table 3.3). The ability to implement action plans seemed to depend on several factors, three of which were age of settlement, levels of household income, and perceived absence of collaboration between the community and administration. Despite being an older settlement with residents being descendants of a common ancestor, Mahindi village was one of the villages with the lowest ability to implement most of their planned actions. The cause of this was not clear but seemed to relate to the quality of leadership in the village.

3.4.2 Community Participation

The participation of communities in generating data on their agroecosystem, analyzing the data, and then using them to make action plans was perceived to be a gainful exercise by the communities and provided an entry point to the communities for the researchers. The most important exercises in this regard were the mapping exercises, historical background, and transect walks. However, the first meeting with the communities was critical since important decisions were made at this point, and community perceptions of the process were formed at this point. If the initial meeting fails to attract representatives of the major stakeholder groups, serious biases to the process may result. Because the researchers had little or no initial contact, it is difficult to provide descriptions and agendas that would ensure the participation of all stakeholders. In this research, divisional government department officials were included in the process to provide insights into the possible stakeholder groups and to ensure stakeholder participation in the initial meetings.

Another constraint to the participatory process is the tendency by communities to bias themselves toward the perceived interests of the researcher. Because of this, participants provide what they consider "correct answers," resulting in tautological biases. This can be minimized in several ways. The first is by providing the communities with a succinct description of the action research process at the beginning.

Another is by avoiding focused discussions on single-discipline issues during the initial stages of the process. The initial meetings should focus mainly on the overall process, the methods, and expected outcomes. Problem statements and problem analyses should therefore be done after the data-gathering and analysis steps. The other factors that reduce tautological biases are the use of a multidisciplinary team of facilitators and the frequency of meetings with communities during the initial process. Frequent meetings imparts to the communities the sense that the process is ongoing and continuous and that the focus is on the communities' real concerns and their agenda. As the engagement between researchers and communities continues, the communities learn that it is their perspective—rather than their attempts to influence researchers' perceptions—that is important. As a result, responses become more detailed and rationally consistent with time. This controverts the perception that quality participatory assessments can be done rapidly.

3.4.3 Systemic Descriptions

In general, human population was perceived as having increased, as did land under cultivation, while soil fertility and farm sizes were seen as declining. Crop and human diseases were reported as having increased. Declining tree cover and the accompanying decline in firewood availability were also reported as an important trend in most of the villages. While these perceptions were confluent with reports from agencies and institutions in the area, the descriptions of what were perceived to be the causes were sometimes divergent. The communities attributed population increases solely to changes in the sociocultural dynamics and traditions as opposed to increased child survival and life expectancy. Increases in land under cultivation and declines in farm sizes were attributed solely to population increases, with no mention of the inability to increase yields per land unit and the cultural practice of land subdivision.

Although the criteria for wealth and well-being were based mostly on material possessions, diet, ability to educate children, personal hygiene, and health status of family were also mentioned. Having an off-farm income was an important criterion in most of the villages. The value of this is likely to be in both the availability of cash income and not being solely dependent on farm income, thus reducing risk. The type of houses ranged widely from grass thatched huts to permanent buildings, but the value of permanent houses seemed to be greater as a social status symbol rather than the direct utility obtained. The fact that most villages indicated that a majority of the people were average in wealth and well-being is more likely a reflection of the sociocultural inhibition of the communities against public discussion of wealth and poverty.

Surprisingly, all communities except Gikabu described malaria as one of the most important diseases in terms of both prevalence and severity. Kiambu district, in the highlands, is not considered climatically suitable for sustained transmission of malaria. SSIs with clinical and medical officers in the health facilities used by these communities confirmed that a diagnosis of malaria was made in the majority of cases exhibiting fever and headache. Furthermore, three of the communities perceived a relationship between increases in the mosquito population and the prevalence of malaria. It remains to be confirmed that there were significant changes in the vector dynamics and infection rates in those recent years.

3.4.4 Problem Analyses and Action Plans

Many of the concerns and problems of the community were related to the poor state of the infrastructure. Poor human health was attributed mostly to the lack of an accessible and functional health care system, while low farm productivity was linked to a run-down extension system. Because of this, many of the solutions proposed were mainly development of infrastructure. Although the communities were aware that many of their goals were to provide common goods that should be provided for through a taxation system, there was a concern that this system was unreliable and not sensitive to local needs.

3.4.5 Monitoring and Evaluation

Communities were able to design and conduct participatory monitoring and evaluation programs for their agroecosystems. This supports the view that the combination of approaches used in this project as well as the health language were well understood by the communities. Details of the methods used in developing indicators used in this process are described in Chapter 6. The self-organized intervillage evaluation meetings by communities in the six ISSs are an indication of the success of the action research process. It also indicates that communities valued the process of monitoring and evaluation, both as a source of inspiration and motivation and as providing support for their decision-making processes. The main difficulty was the cultural inhibitions in the community against public discussion of certain topics. This reduced the usefulness of the monitoring-and-evaluation exercise. The other potential difficulty was that community-driven indicators required a complementary assessment of the researcher-proposed indicators to provide sufficient information on which decisions could be based. Because communities were unable to handle the numerical methods required, they therefore had to depend on external help. The question is how this can be structured to sustain the process.

REFERENCES

Argyris, C., and Schon, D.A. (1991). Participatory action research and action science compared. In *Participatory Action Research.* Whyte, W.F., ed. Newbury Park, CA: Sage, pp. 85–96.

Burkey, S. (1993). Self reliant participatory development. In *People First: A Guide to Self-Reliant Participatory Rural Development.* Burkey, S., ed. Atlantic Highlands, NJ: Zed Books, pp. 40–70.

Cabanero-Verzosa, C., Johnston, C.M., and Kayode, O. (1993). Using focus groups to develop and promote an improved weaning food product. In *Rapid Appraisal Methods.* Kumar, K., ed. London: Intermediate Technology, pp. 94–111.

Carr, W., and Kemmis, S. (1986). *Becoming Critical: Education, Knowledge and Action Research.* Falmer Press, London, p. 249.

Catley, A. (1999). *Methods on the Move: A Review of Veterinary Uses of Participatory Approaches and Methods Focusing on Experiences in Dryland Africa.* International Institute for Environment and Development, London, p. 100.

Chambers, R. (1994). The origins and practice of participatory rural appraisal. *World Development* 22: 953–969.

Chataway, C.J. (1997). An examination of the constraints on mutual inquiry in a participatory action research project. *Journal of Social Issues* 53: 747–765.

Checkland, P., and Scholes, J. (1990). *Soft Systems Methodology in Action*. Wiley, Chichester, U.K. 329 pp.

D'Arcy, D.C. (1990). *The Community's Toolbox: The Idea, Methods and Tools for Participatory Assessment, Monitoring and Evaluation in Community Forestry*. Food and Agricultural Organization, Rome.

DePoy, E., Hartman, A., and Haslett, D. (1999). Critical action research: A model for social work knowing. *Social Work* 44: 560–569.

Eden, C., Jones, S., and Sims, D. (1983). *Messing About in Problems: An Informal Structured Approach to their Identification and Management*. Pergamon Press, Oxford, U.K., p. 124.

Elden, M., and Levin, M. (1991). Cogenerative learning: Bringing participation into action research. In *Participatory Action Research*. Whyte, W.F., ed. Newbury Park, CA: Sage, pp. 127–142.

Freire, P. (1968). *Pedagogy of the Oppressed*. Seabury Press, New York, p. 186.

Glaser, B., and Strauss, A. (1967). *The Discovery of Grounded Theory: Strategies for Qualitative Research*. Aldine, New York.

Grandin, B. (1988). *Wealth Ranking in Smallholder Communities: A Field Manual*. Intermediate Technology, London, p. 49.

Greenwood, D.J., Whyte, W.F., and Harkavy, I. (1993). Participatory action research as a process and as a goal. *Human Relations* 46: 175–192.

Ison, R.L. (1993). Changing community attitudes. *Rangeland Journal* 15: 154–166.

Jiggins, J. (1995). Problem led R&D. In *Grassland into the 21st Century*. Pollot, G.E., ed. British Grassland Society Occasional Symposium, 29, Gloucestershire, U.K., pp. 337–345.

Kabutha, C., Thomas-Slayter, B.P., and Ford, R. (1993). Participatory rural appraisal: A case study from Kenya. In *Rapid Appraisal Methods*. Kumar, K., ed. London: Intermediate Technology, pp. 176–211.

Kumar, K. (1993). *An Overview of Rapid Appraisal Methods in Development Settings. Rapid Appraisal Methods*. Intermediate Technology, London, pp. 8–22.

Puccia, C.J., and Levins, R. (1985). *Qualitative Modeling of Complex Systems*. Harvard University Press, Cambridge, MA. 259 pp.

Ridgley, M., and Lumpkin, C.A. (2000). The bi-polar resource-allocation problem under uncertainty and conflict: A general methodology for the public decision-maker. *Journal of Environmental Management* 59: 89–105.

Rietbergen-McCracken, J., and Narayan, D. (1998). *Participation and Social Assessment: Tools and Techniques*. World Bank, Washington, DC.

Theis, J., and Grady, H. (1991). *Participatory Rapid Appraisal for Community Development: A Training Manual Based on Experience in the Middle East and North Africa*. Intermediate Technology, London, p. 150.

Webb, G. (1996). Becoming critical of action research for development. In: *New Directions in Action Research*. Zuber-Skerrit, O., ed. London: Falmer Press, pp. 137–161.

Webber, L.M., and Ison, R.L. (1995). Participatory rural appraisal design: Conceptual and process issues. *Agricultural Systems* 47: 107–131.

Whyte, W.F. (1991). *Participatory Action Research*. Sage, Newbury Park, CA, p. 247.

Winter, R. (1996). Some principles and procedures for the conduct of action research. In *New Directions in Action Research*. Zuber-Skerrit, O., ed. London: Falmer Press. pp. 13–27.

Zuber-Skerritt, O. (1996). Introduction: New Directions in Action Research. In: *New Directions in Action Research*. Zuber-Skerrit, O., ed. Falmer Press, London, pp. 3–9.

4 Use of Systems Theory, Directed Graphs, and Pulse Process Models in an Adaptive Approach to Agroecosystem Health and Sustainability

4.1 INTRODUCTION

Attempts to understand the interrelationships between, on the one hand, goals and objectives of communities living in an agroecosystem and, on the other hand, their planned actions, stated needs, and concerns require the understanding of a complex system. Such a system involves many variables interacting with each other in a dynamic process. Furthermore, the definition of these variables and their relationships depend on how the communities perceive their world. In attempting to model such a complex system, one faces a trade-off between the accuracy of the model's predictions and the ability to obtain the detailed information needed to build the model (Roberts and Brown, 1975).

A *system*, better referred to as a *holon* to distinguish it from a real-world assemblage of structures and functions, is a representation of a situation and consists of an assembly of elements linked in such a way that they form an organized whole (Flood and Carson, 1993). An element is a representation of some phenomena by a noun or a noun phrase. Links between elements represent a relationship between them. A relationship can be said to exist between two elements if the behavior of one is influenced or controlled by the other (Flood and Carson, 1993). *Behavior* refers to changes in one or more important attributes of an element. *Systems thinking* involves formulating a holon and then using it to find out about, gain insight into, or engineer a part of the perceived world.

The difficulty in formulating a holon to study the interrelationships among community values, community goals, planned actions, and perceived problems arises from three predicaments. The first is that values, goals, and problems are socially constructed based on the perspectives of the stakeholders, and these are sometimes divergent or conflicting (Ison et al., 1997). No one such perspective is sufficient or complete, and none can be said to be right or wrong. Furthermore, problems and concerns in the agroecosystem are often part of what has been referred to as a mess

(Ackoff, 1980). A *mess* is a complex of interrelated problems in which there is no common agreement about the nature of the problems or potential solutions.

The second predicament stems from the fact that many of the relationships between elements in the model reflect human intentions (Caws, 1988), many of which are characterized by a high degree of uncertainty. The third predicament is that information and knowledge needed to build the model depend on human experience. Methods for eliciting experience-based knowledge are characterized by a high degree of subjectivity. The question of how to analyze and interpret community values, goals, and objectives in an agroecosystem is therefore one of how to formulate a problem holon as a composite of all stakeholder perspectives on the problem situation. Such a problem holon must be a problem-determined system (rather than a system-determined problem) that is a sociocultural construct based on the community's perception of biophysical phenomena (Ison et al., 1997).

One way in which a problem-determined holon of an agroecosystem can be derived is by generating a cognitive map of the community's assertions with regard to their collective values, goals, and problems. A cognitive map is a representation of people's assertions about a specified domain. It is derived by depicting how people think an action will achieve their objectives (based on how they understand the world to work) in a graphical form in which concepts are connected to each other by lines and arrows (Ridgley and Lumpkin, 2000). The concepts are represented as words or phrases referring to actions, contexts, quality, or quantities of things in the physical world. The connections reflect relationships thought to exist between the connected concepts. Such relationships can be cause and effect, precedence, or even affinity. Depending on their characteristics, the resulting depictions are variously referred to as cognitive maps, influence diagrams, or directed graphs (digraphs) (Ridgley and Lumpkin, 2000).

The usefulness of cognitive maps depends on two questions (Axelrod, 1976a): (1) Do processes in the modeled domain occur in accordance with the laws of cognitive maps? (2) If they do, is it possible to measure accurately assertions and beliefs of a community in such a way that a model can be applied? Several techniques for eliciting people's assertions have been applied (Axelrod, 1976b), including questionnaire surveys and open-ended interviews. To elicit assertions on factors influencing agroecosystem health and sustainability from communities, the methods should satisfy three requirements. First, the derivation should not require a priori specification of the concepts a particular community may use in its cognitive map. Second, the options, goals, ultimate utility, and relevant intervening concepts should all be included in the cognitive map for it to be useful in evaluating different management options (Axelrod, 1976b). Last, the map should be an accurate representation of the collective assertions (and relationships among them) of the community. Such a cognitive map is better perceived as a signed directed graph, simply known as a digraph (Axelrod, 1976a), with points representing each of the named concepts and arrows representing the relationships between concepts. The arrows are drawn from the "cause" variable to the "effect" variable, with either a positive sign to indicate a direct (or positive) relationship or a minus sign to indicate an inverse (or negative) relationship.

Visual inspection is not a reliable way of analyzing digraphs. A mathematical framework is essential to identify the underlying properties of the digraphs and to enable comparisons between graphs (Sorensen, 1978). There are several mathematical approaches for analyzing signed digraphs based mostly on graph theory, matrix algebra, and discrete and dynamic system models (Harary et al., 1965). The approaches fall into two broad categories: arithmetic and geometric (Roberts, 1976b).

The aim of geometric analysis is usually to analyze the structure, shape, and patterns that may impart important characteristics to the system. A typical geometric conclusion is that some variable will grow exponentially or that some other variable will oscillate in value. The numerical levels reached are not considered important in such predictions (Roberts, 1976b). Geometric analysis of a signed digraph includes (1) tracing out the different causal paths (Axelrod, 1976a), (2) identification of feedback loops (Roberts, 1976b), (3) detection of path imbalance (Nozicka et al., 1976), (4) assessment of stability (Roberts, 1976a), (5) calculation of the strong components, (6) assessment of connectedness (Roberts, 1976b), and (7) assessment of the effects of different strategies (a change in the structure of the system) on system characteristics (Roberts, 1976a).

Arithmetic analyses proceed from the perception of the signed digraph as a dynamic system in which an element obtains a given value with each unit change in time (or space) of another. The values obtained depend on previous changes in other variables. The simplest assumption about how changes of value are propagated through the system is the so-called pulse process (Roberts, 1971). By assuming that change in values in the model follows a specified change-of-value process (such as the pulse process), (1) stability can be assessed even for path-imbalanced digraphs, (2) the effect of outside events on the system can be studied, and (3) forecasts can be made. Roberts (1976a) cautioned that results from arithmetic analyses should be regarded as suggestive and verified by further analysis since digraphs—as models of a complex system—are not precisely correct due to oversimplifications made in the modeling process.

This chapter describes the formulation of a problem-determined holon for an agroecosystem and its analysis using graph theory and dynamic modeling techniques. The overall objective was to gain an insight into the communities' definition of health and to identify the factors they considered to be the most influential in terms of the health and sustainability of their agroecosystems. This analytic framework served as a basis for selecting indicators and in interpreting them. Specifically, the objectives were (1) to assess how communities in the agroecosystem perceived the interrelationships between problems, goals, values, and other factors; (2) to evaluate what the communities perceived to be the overall benefits of various agroecosystem management strategies; (3) to determine what would be the most relevant measures of change in the problem situation; and (4) to find what would be the long-term effects of various strategies and management policies, assuming that the communities' assertions were reasonably accurate depictions of the problem situation.

4.2 PROCESS AND METHODS

Cognitive maps (also known as loop models, influence or spaghetti diagrams) were defined as models that portrayed ideas, beliefs, and attitudes and their relationship to one another in a form amenable to study and analysis (Eden et al., 1983; Puccia and

Levins, 1985; Ridgley and Lumpkin, 2000). Cognitive maps were developed, one for each intensive study site (ISS), in 1-day participatory workshops, using principles of participatory mapping described in Chapter 3. The maps were analyzed using graph theory as described by Harary et al. (1965), Jeffries (1974), Roberts and Brown (1975), Roberts (1976a, 1976b), Perry (1983), Puccia and Levins (1985), Klee (1989), Ridgley and Lumpkin (2000), and Bang-Jensen and Gutin (2001).

4.2.1 Participatory Cognitive Mapping

Cognitive maps, in the form of signed directed graphs (digraphs), were constructed for each ISS. These mapping activities were carried out in October and November 1997, subsequent to the initial village workshops. Details of the selection of study sites are provided in Chapter 2. A 1-day workshop was held in each study site. Each household in the study site was represented by at least one person. Although workshop participants from the ISS communities were not necessarily experts in any relevant technical discipline, they were considered "lay" experts (Roberts, 1976a) due to their unique experiential knowledge of the agroecosystem. Local participants were taken to be "synthetic experts" (Dalkey, 1969).

To facilitate group discussions and to provide opportunities for each local participant to give an opinion, the local participants were divided into groups of 6–10. The number (ranging from 4 to 10) of groups depended on the number of participants and therefore the size of the village. A facilitator and a recorder were provided for each of the groups. Facilitators consisted of researchers and divisional team members as described in Chapter 2. Each group was asked to discuss how various problems and concerns in the study site interacted with each other, thus precipitating changes in the health and sustainability of the agroecosystem. A whiteboard, index cards, and large sheets of paper were used to plot the graphs. Each group was shown, using an abstract example, of how they could represent their views in the form of a digraph using the materials provided. Participants were asked to record the concepts on index cards (making it easier to move concepts in a diagram) or directly on a whiteboard. The concepts were then to be linked using the rules described for cognitive maps and signed digraphs. Each group presented its diagram to the rest of the workshop participants. Diagrams were compared and contrasted and a composite diagram developed. This composite diagram included only those concepts and relationships in which there was consensus about their existence. The rationale for this was that collective action was likely to follow only if consensus existed. Further, consensus was assumed to indicate a collective agreement that the concepts and relationships operated in the manner depicted.

Participants described relationships among concepts in terms of the direction of influence (for example, A influences B), the sign (positive if positively correlated and negative if negatively correlated), as well as the perceived impact on the system (positive if beneficial and negative if detrimental). In the cognitive map, correlations were denoted by the line form (solid if positive and dashed if negative). The impact was denoted by the color; red arrows denoted negative impact, while blue lines denoted positive impact. A solid red arrow, for example, represented a positive correlation with a negative impact on the agroecosystem. Conversely, a dashed blue line represented a negative correlation with a positive impact.

At all the study sites, participants began by listing categories of concepts needed to explain the relationships between, on the one hand, agroecosystem problems and concerns and, on the other, its health and sustainability. A metaphor in the local language was used to equate categories of related concepts to pots and the thought process as cooking. Categories, and eventually the concepts themselves, were generated using declarative statements of the form, "You cannot cook (think about) x without (including the concept of) y." Concepts belonging to the same "pot"—those seen to be related in some ways—were circled if on a chalkboard or put in one pile if on cards. Relationships between pots were then added to the diagram, followed by relationships within.

4.2.2 Geometric Analyses

A *signed digraph* $D = (V, A)$ was defined as consisting of a set (V) of points (v_1, v_2, …, v_n) called vertices and another set (A) of dimensions $n \times n$ called the adjacency matrix (Figure 4.1). The adjacency matrix of a digraph $D = (V, A)$ consists of elements a_{ij}, where $a_{ij} = 1$ if the arc (v_i, v_j) exists and 0 if the arc (v_i, v_j) does not exist, with i and $j = \{1, 2, 3, \ldots, n\}$. The in-degree of a vertex (v_i) is the sum of the column (i) in the adjacency matrix corresponding to that vertex. Conversely, the out-degree of a vertex (v_i) is the sum of the row (i) in the adjacency matrix corresponding to that vertex. The sum of the in-degree and the out-degree of a vertex is the total degree (td) and is a measure of the cognitive centrality of the vertex (Nozicka et al., 1976). A vertex with an in-degree of 0 was described as a source, while one with an out-degree of 0 was described as a sink.

A *path* was defined as a sequence of distinct vertices (v_1, v_2, …, v_t) connected by arcs such that for all $i = \{1, 2, \ldots, t\}$ there is an arc (v_i, v_{i+1}). The sign (or effect) of a path was the product of the signs of its arcs, and the length of a path was the number of arcs in it. The impact of a path from vertex v_i to another vertex v_j was calculated as the effect of the path multiplied by the sign of vertex v_j. The sign of a vertex was

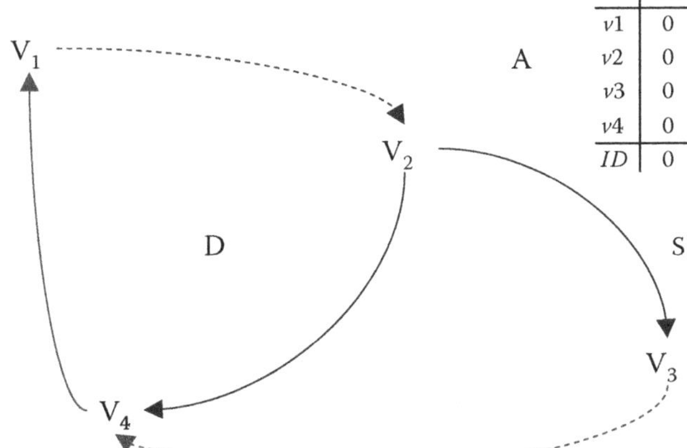

A

	v1	v2	v3	v4	OD
v1	0	1	0	1	2
v2	0	0	1	1	2
v3	0	0	0	1	1
v4	0	0	0	0	0
ID	0	1	1	3	

Sgn(A)

0	−1	0	0
0	0	1	1
0	0	0	−1
1	0	0	0

FIGURE 4.1 Example of a digraph and its adjacency (A) and signed adjacency (sgn(A)) matrices. See CD for color image.

positive if all positive-effect arcs leading to it had a positive impact and negative if otherwise. The sign of a source vertex was the sum of the impacts of all arcs leading from it. In contrast to a path, a *cycle* was defined as a sequence of vertices (v_1, v_2, …, v_t) such that for all $i = \{1, 2, \ldots, t\}$ there is an arc (v_i, $v_{(i+1)}$), and where $v_1 = v_t$, while all other vertices are distinct. The sign, length, and impact of a cycle were as defined for paths. The diagonal elements (a_{ii}) of the matrix A^t gave the number of cycles and closed walks from a given vertex (v_i). The off-diagonal elements gave the number of walks and paths from one vertex (v_i) to another (v_j). A walk was similar to a path with the exception that the vertices forming the sequence were not distinct.

The total effect (*TE*) of a vertex (v_i) on another vertex (v_j) is the sum of the effects of all the paths from v_i to v_j. If all such effects are positive, then the total effect is positive (+); if all are negative, the total effect is negative (–); if two or more paths of the same length have opposite effects, the sum is indeterminate (#), and if all the paths with opposite effects are of different lengths, the sum is ambivalent (±). A digraph with at least one indeterminate or ambivalent total effect is said to be path imbalanced. One that has no indeterminate or ambivalent total effect is path balanced. The signed adjacency matrix (also called the incidence matrix, direct effects matrix, or valency matrix) is used to compute the total effect. The impact of vertex v_i on another vertex v_j is calculated as the total effect of v_i on v_j multiplied by the sign of vertex v_j.

The reachability matrix R is a square $n \times n$ matrix with elements r_{ij} that are 1 if v_j is reachable from v_i and 0 if otherwise. By definition, each element is reachable from itself, such that $r_{ii} = 1$ for all i. The reachability matrix can be computed from the adjacency matrix using the formula $R = B[(I + A)^{n-1}]$. B is a Boolean function where $B(x) = 0$ if $x = 0$, and $B(x) = 1$ if $x > 0$. I is the identity matrix. The digraph $D = (V, A)$ is said to be strongly connected (i.e., for every pair of vertices v_i and v_j, v_i is reachable from v_j and v_j is reachable from v_i) if and only if $R = J$, the matrix of all 1's. D is unilaterally connected (i.e., for every pair of vertices v_i and v_j, v_i is reachable from v_j or v_j is reachable from v_i) if and only if $B[R + R'] = J$. The strong component (i.e., a subdigraph of D where all the vertices are maximally connected) to which a vertex (v_i) is a member is given by the entries of 1 in the ith row (or column) of the elementwise product of R and R'. The number of elements in each strong component is given by the main diagonal elements of R^2.

4.2.3 Pulse Process Models

A *weighted digraph* is one in which each arc (v_i, v_j) is associated with a weight (a_{ij}). The signed adjacency matrix (in this case referred to as a weighted adjacency matrix) of a weighted digraph therefore consists of the signed weights (a_{ij}) of all the arcs (v_i, v_j) in the digraphs and is 0 if the arc does not exist. Under the pulse process, an arc (v_i, v_j) was interpreted as implying that when the value of v_i is increased by one unit at a discrete step t in time or space, v_j would increase (or decrease depending on the sign of a_{ii}) by a_{ij} units at step $t + 1$. Initially, the arcs in each digraph were considered to be equal in weight and length. The models therefore assumed that a pulse in vertex v_i at time t was related in a linear fashion to the pulse in v_j at time

$t + 1$ if there was an arc (v_i, v_j) in the digraph. The value (v_{it}) of vertex v_i at time t was calculated as:

$$v_{it} = v_{i(t-1)} + P^{o}_{i(t-1)} + \sum_{j=1}^{n} \text{sgn}(v_j, v_i) P_{j(t-1)}$$

$P^0_{i(t-1)}$ is a vector of external pulses or change in vertices $v_1, v_2, \ldots, v_n$ at step $(t - 1)$; $\text{sgn}(v_i, v_j)$ is the sign of arc (v_i, v_j); $P_{j(t-1)}$ is referred to as a pulse and is the jth element of the pulse vector P at the $(t - 1)$th row. P_{jt} is given by the difference $v_{jt} - v_{j(t-1)}$ for $t > 0$ and 0 otherwise.

A pulse process of a signed digraph D was defined by a vector of the starting values at each vertex given by $V_s = \{v_{1s}, v_{2s}, \ldots, v_{ns}\}$ and a vector of the initial pulses at each of the vertices given by $P^0_0 = P_0 = \{P_{10}, P_{20}, \ldots, P_{n0}\}$. Thus, the value at vertex v_i at step $t = 0$ was calculated as $u_{i0} = u_{is} + p_{i0}$.

A pulse process is autonomous if $p^0_i(t) = 0$ for all $t > 0$, that is, no other external pulses are applied after the initial pulse P_0 at step $t = 0$. In an autonomous pulse process in a digraph, $D = (V, A)$, $P_t = (P_0 * A^t)$. Further, a pulse process starting at vertex v_i is described as simple if P_0 has the ith entry equal to 1 and all other entries equal to 0; that is, the system receives the initial pulse from a single vertex. Under a simple autonomous pulse process, a unit pulse is propagated through the system starting at the initial vertex v_i. Under this process, the value of vertex v_i at time t is given by

$$v_{it} = v_{i(t-1)} + \sum_{j=1}^{n} \text{sgn}(v_j, v_i) P_{j(t-1)}$$

From this, it can be shown that in a simple autonomous pulse process starting at vertex v_i, the value at vertex v_j at step t is given by $u_j(t) = u_j(0) + e_{ij}$, where e_{ij} is the i,jth element of a matrix $T = (A + A^2 + \ldots + A^t)$, where A is the weighted adjacency matrix.

The effect of a vertex v_i on another v_j was positive if all pulses at v_j resulting from a simple autonomous pulse at v_i were always positive, ambivalent if they were oscillating, and positive if they were always negative. The impact was calculated as described in the geometric analysis.

Based on the work of Klee (1989), a digraph was described as stable, value (or quasi-) stable, semistable, or unstable under a given pulse process. A digraph was stable under a pulse process if the values at each vertex converged to the origin as $t \rightarrow \infty$. It was described as value stable if the values at each vertex were bounded, that is, there were numbers B_j so that $|v_{jt}| < B_j$ for all j and $0 \leq t \leq \infty$. A digraph was semistable if the values at each vertex changed at a polynomial rather than an exponential rate. It was unstable if the converse was true. A digraph was described as pulse stable under a pulse process if the pulses at each vertex were bounded for $0 \leq t \leq \infty$, that is, $|p_{jt}| < B_j$ for all t. Stability properties of a digraph are related to the eigenvalues of the weighted adjacency matrix. A digraph was stable under all pulse processes if and only if each eigenvalue had a negative real part (Klee, 1989). If all nonzero eigenvalues of A were distinct and at most 1 in magnitude, then the digraph was pulse stable under all simple pulse processes (Roberts and Brown, 1975). A digraph was value stable under all simple pulse processes if it was pulse stable and 1 was not an eigenvalue of D (Roberts and Brown, 1975). A digraph was semistable under all pulse processes if and only if each eigenvalue had a nonpositive real part (Klee, 1989).

4.2.4 Application of System Theory Tools in Villages

Sources in a digraph were seen as representing those factors requiring external intervention. Perceived impacts and expected outputs of community goals were assessed in two ways. The first was through geometric analysis of the cognitive maps, which involved examination of the total impacts of vertices corresponding to each of the goals. The total number of positive impacts was used to rank community goals, and this was compared to the ranking done by communities during the participatory workshops. Presence of indeterminate effects was considered a result of path imbalance. Path imbalances were seen as those relationships in which the outcome could be either negative or positive depending on the weight and time lags placed on the arcs of the various paths linking the vertices. These were considered important as they represented aspects for which trade-offs and balances were critical to the overall outcome of community goals. Presence of ambivalent impacts was seen as an indication of the system's increased amplitude instability.

The second method of assessing the impact of community goals was simple autonomous pulse processes initiated at each of the vertices corresponding to a community goal. The impact was assessed based on $(n - 1)$ iterations, equivalent to the longest path in the digraph. The usefulness of this approach was in assessing the importance of path imbalance in the outcome of community goals. Digraphs in which community goals had only positive impacts were said to be in regenerative spirals. Those in which there was a preponderance of negative impacts were said to be in degenerative spirals.

Two kinds of value-stabilizing strategies were assessed. First was where the signs of arcs in the digraph were changed either individually or as a group. Stabilizing strategies involving the fewest changes were considered the simplest. The other type of stabilizing strategies was where the weights associated with the arcs were altered—with the simplest strategies—those that involved the fewest changes. The importance of assessing value stability was to evaluate the key relationships on which the impacts of community goals were predicated. Existence of many simple stabilizing strategies was considered an indication of increased system inertia. Absence of stabilizing strategies was considered an indication not only of cognitive imbalance but also as possible trajectory stability.

4.3 RESULTS

Three groups of concepts were common to cognitive maps of the six communities. These were problems, outputs, and institutions. For ease of analysis, the common categories were retained, while the rest of the concepts were placed into one general category: system-state (Figure 4.2). The number of concepts depicted in the

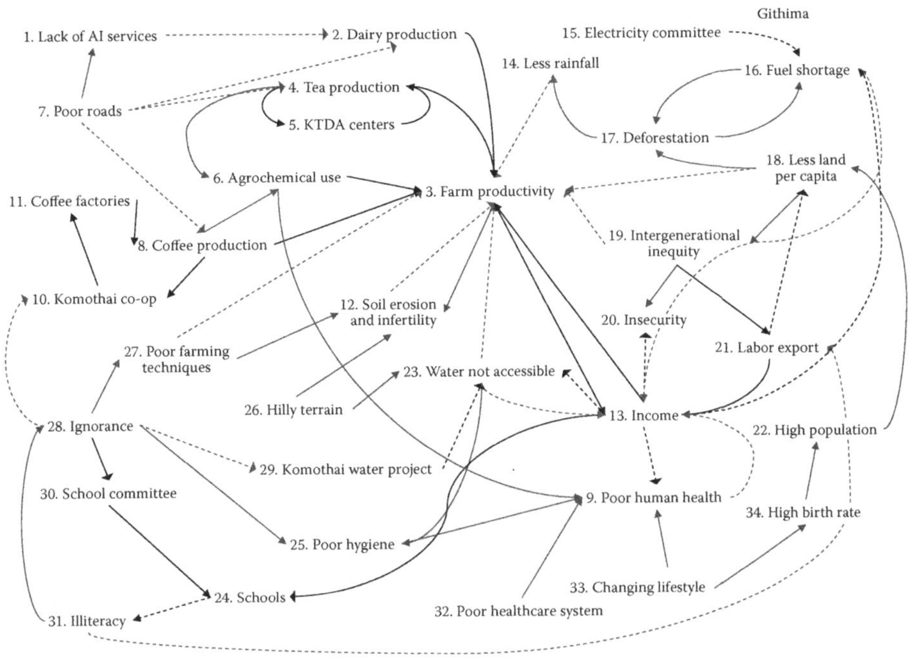

FIGURE 4.2 A cognitive map depicting perception factors influencing agroecosystem health and sustainability in Githima intensive survey site, Kiambu District, Kenya, 1997. AI, artificial insemination. (KTDA = Kenya Tea Development Authority) See CD for color image.

cognitive maps from the different communities was similar. Mahindi had the most (38), while Thiririka and Gitangu had the least (31) (Table 4.1). The cognitive map by the Kiawamagira community had the most (66) arcs, followed by that by Githima (Table 4.1). The cognitive map drawn by the Thiririka community had the lowest average number of relationships per concept (1.5), followed by Mahindi (1.6), and then Gikabu (1.7). Githima and Gitangu had the highest (1.9) number of relationships per concept.

In all villages, relationships with negative impacts were the most preponderant, comprising between 60% and 70% of all the arcs in the digraphs. Mahindi and Thiririka villages had the highest proportion of negative-impact relationships (71.2% and 70.8%, respectively). Mahindi and Gitangu each showed only one institution in their influence diagrams despite having mentioned several of them in the institutional analysis.

4.3.1 Githima

The cognitive map depicting the perceptions of the residents of Githima village is shown in Figure 4.2. Vertex 3, with a total degree of 12, has cognitive centrality. Other vertices with high total degree are 13, 9, and 23 with total degrees of 11, 6, and 6, respectively. Vertex 20 is the only sink (out-degree = 0), while vertices 7, 15, 26, 32, and 33 are sources (in-degree = 0).

TABLE 4.1

A Comparison of the Number of Concepts and Relationships in Cognitive Maps Drawn by Six Communities in Kiambu District, Kenya, Depicting Community Perceptions of Factors Influencing Agroecosystem Health and Sustainability

	Number of Concepts					Number of Arcs	
Village	Total	Problems	Outputs	States	Institutions	Total	% with Negative Effect
Githima	34	8	4	15	7	63	63.5
Gitangu	31	11	4	15	1	59	64.4
Kiawamagira	37	10	4	16	7	66	69.7
Mahindi	38	6	3	28	1	59	71.2
Gikabu	33	10	3	13	7	57	66.7
Thiririka	31	10	3	15	3	48	70.8

The impacts of Githima community's goals, based on a geometric analysis of their cognitive map of factors influencing agroecosystem health and sustainability, are shown in Table 4.2. Roads, knowledge, and illiteracy had indeterminate impacts on most vertices. These result from two imbalanced paths from vertex 6 (agrochemical use) to vertex 13 (income). All goals had negative impacts on agrochemical use. This is because it is a negative vertex but with positive impact on farm productivity.

All goals except roads had a negative impact on vertex 30 (school committee), caused by the positive-impact negative-feedback loop linking it to the negative vertex 28 (ignorance). All goals except artificial insemination (AI) and security had indeterminate impacts on vertex 12 (soil erosion and infertility). The indeterminate impacts of roads, knowledge, and literacy on the soil vertex were due to the path imbalance between vertices 6 and 13. The indeterminate impacts of health and health care on the soil vertex resulted from path imbalance between vertices 13 and 12 (the positive path passes through vertex 16, while the negative one passes through vertex 27).

When arc [6, 9] is negative or absent, the overall positive impacts of community goals increase to 154 with only 16 negative impacts. This results mostly from an increase in the positive impacts of roads and literacy. Removing the arc [8, 6] increases the overall impact of community goals to 134 while reducing negative impacts to 8. Setting arc [13, 24] to either negative or zero reduces positive impacts of community goals to 45 and 73, respectively, while increasing the negative impacts to 60 and 16, respectively. Similarly, inverting or removing the arc [24, 31] results in reduced positive impacts (50 and 78, respectively). Inverting the arc increases negative impacts to 55, but removing the arc reduces negative impacts to 10.

The digraph consists of 25 feedback loops, only 4 of which are negative feedback. The longest of all the feedback loops are of length nine. There are two strong components. The first has two vertices (tea production and tea centers) in a positive-feedback loop. The other strong component includes all the other vertices except AI services, dairy production, roads, electricity committee, security, population, terrain, health care, lifestyle, and birth rate.

TABLE 4.2
Impact of Githima Community's Goals Based on Geometric Analysis

Vertex	Community Goals								
	Artificial Insemination	Roads	Health	Fuel	Security	Water	Knowledge	Literacy	Health Care
1 (AI)	+	+	.	.	.	.	.	.	.
2 (Dairy)	+	+	.	.	.	.	.	.	.
3 (Productivity)	+	±	+	+	.	+	#	#	+
4 (Tea)	.	+	.	.	.	.	.	.	.
5 (Tea centers)	.	+	.	.	.	.	.	.	.
6 (Chemicals)	–	–	–	–	.	–	–	–	–
7 (Roads)	.	+	.	.	.	.	.	.	.
8 (Coffee)	+	#	+	+	.	+	+	+	+
9 (Health)	±	#	+	#	.	±	#	#	+
10 (Co-op)	+	#	+	+	.	+	+	+	+
11 (Factories)	+	#	+	+	.	+	+	+	+
12 (Soil)	±	#	#	#	.	#	#	#	#
13 (Income)	+	#	+	+	.	+	#	#	+
14 (Rainfall)	+	#	+	+	.	+	#	#	+
15 (Electricity)	.	.	.	.	.	.	.	.	.
16 (Fuel)	+	#	+	+	.	+	#	#	+
17 (Forest)	+	#	+	+	.	+	#	#	+
18 (Land size)	+	#	+	+	.	+	#	+	+
19 (Inequity)	+	#	+	+	.	+	#	+	+
20 (Security)	+	#	+	+	+	+	#	#	+
21 (Labor)	+	#	+	+	.	+	#	+	+
22 (Population)	.	.	.	.	.	.	.	.	.
23 (Water)	+	#	+	+	.	+	#	#	+
24 (Schools)	+	#	+	+	.	+	#	#	+
25 (Hygiene)	+	#	+	+	.	+	#	#	+
26 (Terrain)	.	.	.	.	.	.	.	.	.
27 (Techniques)	+	#	+	+	.	+	+	+	+
28 (Ignorance)	+	#	+	+	.	+	+	+	+
29 (Water project)	+	#	+	+	.	+	+	+	+
30 (School committee)	–	#	–	–	.	–	–	–	–
31 (Illiteracy)	+	#	+	+	.	+	#	+	+

(continued on next page)

TABLE 4.2 (continued)
Impact of Githima Community's Goals Based on Geometric Analysis

Vertex		Community Goals								
		Artificial Insemination	Roads	Health	Fuel	Security	Water	Knowledge	Literacy	Health Care
32 (Health care)		.	.	.	.	.	.	.	.	+
33 (Lifestyle)		.	.	.	.	.	.	.	.	.
34 (Birth rate)		.	.	.	.	.	.	.	.	.
Totals										
+	122	21	5	20	19	1	19	6	10	21
–	**15**	2	1	2	2	0	2	2	2	2
Problem ranking[a]		6	4	2	5	7	1	8	3	2
Goal status[b]		0	2	0	0	3	3	0	1	0

+ Positive impact; – negative impact; ± ambivalent; . no impact; # indeterminate

[a] Ranking by communities during the initial village workshops

[b] Goal status as ranked by communities in January 2000 (0, no change; 1, slight improvement; 2, moderate improvement; 3, improved a lot)

The digraph is unstable under all simple autonomous pulse processes if all arcs are assumed to have equal weights and time lags, the highest eigenvalue being 2.26. Simple positive autonomous pulses representing community goals (except security, which is a sink) lead to negative impacts at vertices 6 (agrochemical use), 12 (soil erosion and infertility), and 30 (school committee) (Table 4.3). In addition to these, improved access roads produces ambivalent impacts at vertex 9, while increased knowledge produces ambivalent impacts at most of the other vertices (Figure 4.3). Ambivalent impacts also occur at vertices 18, 19, and 21, resulting from increased literacy.

The arcs with a change in weight that results in changes in the number of positive impacts of community goals are shown in Appendix 2. Of the 193 impacts of community goals, 165 are sensitive to changes in the weights of at least one arc in the digraph (Table 4.3). The only indirect and nonambivalent impacts that are not sensitive to weight changes are those of roads and AI on vertices 2, 4, and 5. Impacts of community goals were most sensitive to increases in the weight of arcs [3, 12] and [12, 3]. Increases in the weight of any one of these arcs increase the number of oscillating impacts of community goals. A weight of 10 resulted in oscillations of all but nine of the impacts of community goals. Of all the arcs, [31, 21] produced the most changes in the impact of community goals when the weight of each was reduced to values below 1 and above 0.

TABLE 4.3
Impact of Githima Community's Goals Based on a Pulse Process Analysis

Vertex	Community Goals								
	Artificial Insemination	Roads	Health	Fuel	Security	Water	Knowledge	Literacy	Health Care
1 (AI)	+[a]	+[a]	.	.	.	.	.	.	.
2 (Dairy)	+[a]	+[a]	.	.	.	.	.	.	.
3 (Productivity)	+	+	+	+	.	+	+	+	+
4 (Tea)	.	+[a]	.	.	.	.	.	.	.
5 (Tea centers)	.	+[a]	.	.	.	.	.	.	.
6 (Chemicals)	−	−	−	−	.	−	±[a]	−	−
7 (Roads)	.	+[a]	.	.	.	.	.	.	.
8 (Coffee)	+	+	+	+	.	+	±[a]	+	+
9 (Health)	+	±[a]	+	+	.	+	+	+	+
10 (Co-op)	+	+	+	+	.	+	±[a]	+	+
11 (Factories)	+	+	+	+	.	+	±[a]	+	+
12 (Soil)	−	−	−	−	.	−	±[a]	±	−
13 (Income)	+	+	+	+	.	+	+	+	+
14 (Rainfall)	+	+	+	+	.	+	+	+	+
15 (Electricity)	.	.	.	.	.	.	.	.	.
16 (Fuel)	+	+	+	+	.	+	+	+	+
17 (Forest)	+	+	+	+	.	+	+	+	+
18 (Land size)	+	+	+	+	.	+	±[a]	±	+
19 (Inequity)	+	+	+	+	.	+	±[a]	±	+
20 (Security)	+	+	+	+	+[a]	+	+	+	+
21 (Labor)	+	+	+	+	.	+	±[a]	±	+
22 (Population)	.	.	.	.	.	.	.	.	.
23 (Water)	+	+	+	+	.	+	+	+	+
24 (Schools)	+	+	+	+	.	+	±[a]	+	+
25 (Hygiene)	+	+	+	+	.	+	+	+	+
26 (Terrain)	.	.	.	.	.	.	.	.	.
27 (Techniques)	+	+	+	+	.	+	±[a]	+	+
28 (Ignorance)	+	+	+	+	.	+	±[a]	+	+
29 (Water project)	+	+	+	+	.	+	±[a]	+	+
30 (School committee)	−	−	−	−	.	−	±[a]	−	−

(continued on next page)

TABLE 4.3 (continued)
Impact of Githima Community's Goals Based on a Pulse Process Analysis

Vertex		Community Goals								
		Artificial Insemination	Roads	Health	Fuel	Security	Water	Knowledge	Literacy	Health Care
31 (Illiteracy)		+	+	+	+	.	+	±[a]	+	+
32 (Health care)		.	.	.	.	.	.	.	.	+[a]
33 (Lifestyle)		.	.	.	.	.	.	.	.	.
34 (Birth rate)		.	.	.	.	.	.	.	.	.
Totals										
+	154	22	24	20	20	1	20	9	17	21
–	20	3	3	3	3	0	3	0	2	3

+ Positive impact; – negative impact; ± ambivalent; . no impact

[a] Impacts that are not sensitive to weight changes

4.3.2 Gitangu

Figure 4.4 is a cognitive map depicting Gitangu community's perception of factors influencing agroecosystem health and sustainability. Vertices 6 and 11 had cognitive centrality, each with a total degree of 10, followed by vertices 4, 3, and 1, which had total degrees of 9, 8, and 7, respectively. The digraph has no sinks, but six of the vertices (5, 10, 17, 18, 20, and 21) are sources.

The impacts of community goals—based on a geometric analysis—are shown in Table 4.4. All goals except health, security, and secondary schools had indeterminate impacts on vertices 11, 14, 15, and 16. This results from the presence of three equal-length (three arcs in each) paths from vertex 6 to vertex 11, with two positive in effect and one negative. Ambivalent impacts occur at vertices 13, 16, and 27, indicating the presence of counteracting paths. The total number of positive impacts of community goals increases to 147 if arc [12, 27] is inverted and to 137 if it is removed. In both cases, the negative impacts reduce to zero. Positive impacts also increase if arc [27, 11] is removed (136) or inverted (128), but the negative impacts remain 10. Removing arc [6, 12] increases positive impacts (to 113), but negative impacts are reduced to 1. The number of positive impacts of community goals reduces to 70 or less if any one of arcs [13, 4], [2, 3], [3, 8], and [3, 7] are inverted.

The digraph is unstable under all simple autonomous pulse processes if all arcs are taken as having unit weight and time lag. The largest eigenvalue is 2.29. The impact of community goals under a simple autonomous pulse process is shown in Table 4.5. All impacts are positive or ambivalent except at vertex 27, where eight of the goals have negative impact. Most (165/193) of the impacts of community goals are

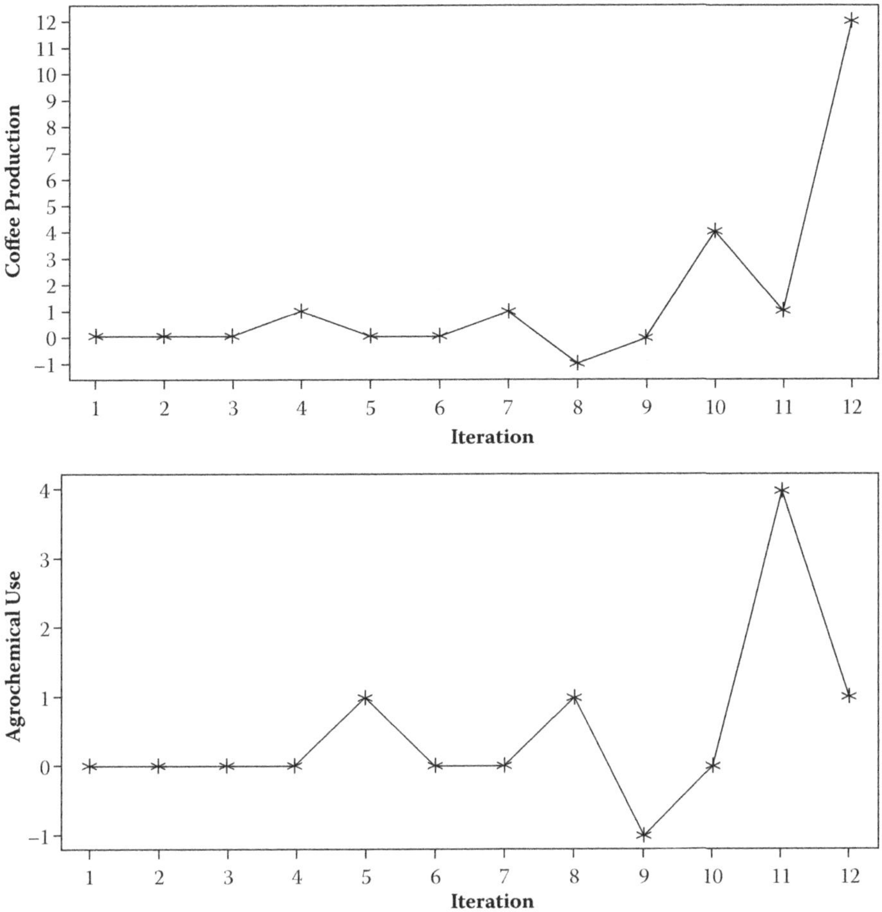

FIGURE 4.3 Oscillating impacts of knowledge at vertices 6 (agrochemical use) and 7 (coffee production) in a pulse process analysis of Githima digraph.

sensitive to increase in the weight of at least one arc in the digraph (Table 4.5). Of the 28 impacts that are not sensitive to increases in the weight of arcs, only 8 are indirect and nonambivalent. The ambivalent impacts of soil fertility on vertices 12, 27, and 31 stabilize as a result of increases in the weights of some of the arcs in the digraph.

The digraph consists of two main (with more than two vertices) strong components. The first strong component comprises vertices 3, 4, 7, 8, and 13 linked by two positive- and one negative-feedback loops. The second consists of vertices 6, 11, 12, 14, 15, 16, 27, 28, and 31 joined into 15 feedback loops, 3 of which are negative. The first strong component is pulse stable. Inverting any one of arcs [3, 7], [4, 3], and [7, 13] makes this strong component value stable under all simple autonomous pulse processes. The second strong component is unstable. Among the simplest strategies that produce value stability are (1) removal of arc [11, 16] accompanied by inversion of arc [15, 11] and (2) removal of arc [14, 11] accompanied by inversion of arc [15, 11].

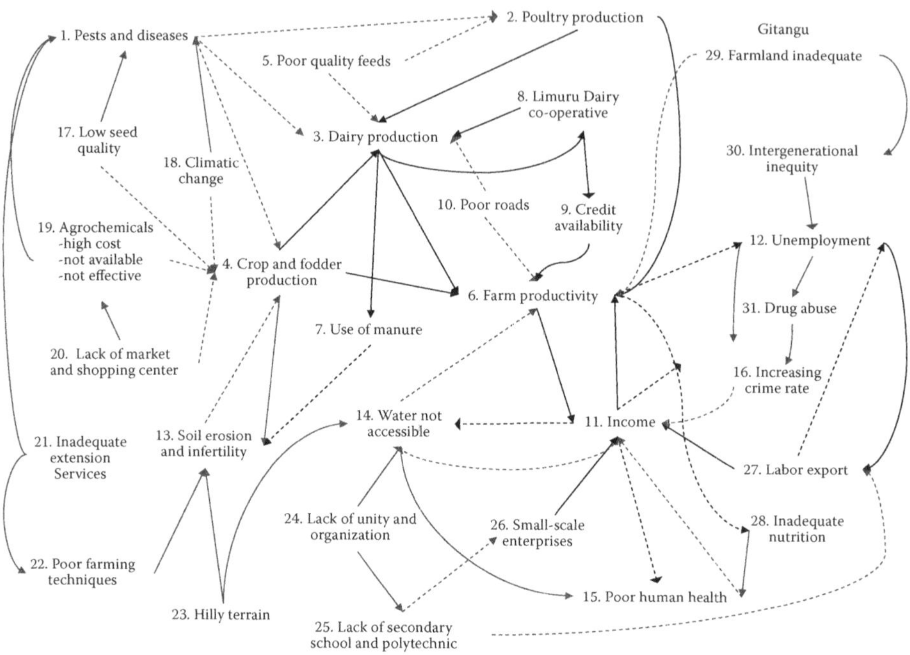

FIGURE 4.4 A cognitive map depicting perception of factors influencing agroecosystem health and sustainability in Gitangu intensive survey site, Kiambu District, 1997. See CD for color image and key.

TABLE 4.4
Impact of Gitangu Community's Goals Based on Geometric Analysis

	Community Goals										
Vertices	Pests and Diseases	Feed Quality	Roads	Employment	Soil Productivity	Water Availability	Human Health	Crime Rate	Market	Extension	Secondary Schools
1 (Diseases)	+	.	.	.	.	.	.	.	+	+	.
2 (Poultry)	+	+	.	.	.	.	.	.	+	+	.
3 (Dairy)	+	+	+	.	+	.	.	.	+	+	.
4 (Food crops)	+	+	+	.	+	.	.	.	+	+	.
5 (Feed quality)	.	+	.	.	.	.	.	.	.	.	.
6 (Productivity)	+	+	+	#	+	+	+	+	+	+	+
7 (Manure)	+	+	+	.	+	.	.	.	+	+	.

TABLE 4.4 (continued)
Impact of Gitangu Community's Goals Based on Geometric Analysis

	Community Goals										
Vertices	Pests and Diseases	Feed Quality	Roads	Employment	Soil Productivity	Water Availability	Human Health	Crime Rate	Market	Extension	Secondary Schools
8 (Co-op)	+	+	+	.	+	.	.	.	+	+	.
9 (Credit)	+	+	+	.	+	.	.	.	+	+	.
10 (Roads)	.	.	+	.	.	.	.	.	.	.	.
11 (Income)	#	#	#	#	#	#	+	+	#	#	+
12 (Employment)	+	+	+	+	+	+	+	+	+	+	+
13 (Soil)	±	+	+	.	+	.	.	.	#	±	.
14 (Water)	#	#	#	#	#	+	+	+	#	#	+
15 (Health)	#	#	#	#	#	#	+	+	#	#	+
16 (Security)	#	#	#	±	#	#	+	+	#	#	+
17 (Seed quality)	.	.	.	.	.	.	.	.	.	.	.
18 (Climate)	.	.	.	.	.	.	.	.	.	.	.
19 (Chemicals)	.	.	.	.	.	.	.	.	+	.	.
20 (Market)	.	.	.	.	.	.	.	.	+	.	.
21 (Extension)	.	.	.	.	.	.	.	.	.	+	.
22 (Techniques)	.	.	.	.	.	.	.	.	.	+	.
23 (Terrain)	.	.	.	.	.	.	.	.	.	.	.
24 (Organization)	.	.	.	.	.	.	.	.	.	.	.
25 (School)	.	.	.	.	.	.	.	.	.	.	+
26 (Enterprises)	.	.	.	.	.	.	.	.	.	.	+
27 (Labor)	–	–	–	–	–	–	–	–	–	–	±
28 (Nutrition)	+	+	+	#	+	+	+	+	+	+	+
29 (Farmland)	.	.	.	.	.	.	.	.	.	.	.
30 (Inequity)	.	.	.	.	.	.	.	.	.	.	.
31 (Drug abuse)	+	+	+	+	+	+	+	+	+	+	+
Totals											
+ 103	11	12	11	2	10	5	8	8	13	13	10
– 10	1	1	1	1	1	1	1	1	1	1	0
Problem ranking	6	7	2	4	8	1	3	4	9	10	5
Goal status	0	1	0	0	1	1	0	3	0	3	0

+ Positive impact; – negative impact; ± ambivalent; . no impact; # indeterminate

TABLE 4.5
Impact of Gitangu Community's Goals Based on Pulse Process Analysis

	Community Goals										
Vertex	Pests and Diseases	Feed Quality	Roads	Employment	Soil Fertility	Water Availability	Human Health	Security	Market	Extension	Secondary Schools
1 (Diseases)	+[a]	.	.	.	.	.	.	.	+[a]	+[a]	.
2 (Poultry)	+[a]	+[a]	.	.	.	.	.	.	+[a]	+[a]	.
3 (Dairy)	+	+	+	.	+	.	.	.	+	+	.
4 (Food crops)	±[a]	+	+	.	±[a]	.	.	.	±[a]	±[a]	.
5 (Feed quality)	.	+[a]	.	.	.	.	.	.	.	.	.
6 (Productivity)	+	+	+	+	+	+	+	+	+	+	+
7 (Manure)	+	+	+	.	+	.	.	.	+	+	.
8 (Co-op)	+	+	+	.	+	.	.	.	+	+	.
9 (Credit)	+	+	+	.	+	.	.	.	+	+	.
10 (Roads)	.	.	+[a]	.	.	.	.	.	.	.	.
11 (Income)	+	+	+	+	+	+	+	+	+	+	+
12 (Employment)	+	+	+	±[a]	±	+	+	+	+	+	+
13 (Soil)	±[a]	+	+	.	±[a]	.	.	.	±[a]	±[a]	.
14 (Water)	+	+	+	+	+	+	+	+	+	+	+
15 (Health)	+	+	+	+	+	+	+	+	+	+	+
16 (Security)	+	+	+	±	+	+	+	+	+	+	+
17 (Seed quality)	.	.	.	.	.	.	.	.	.	.	.
18 (Climate)	.	.	.	.	.	.	.	.	.	.	.
19 (Chemicals)	.	.	.	.	.	.	.	.	+[a]	.	.
20 (Market)	.	.	.	.	.	.	.	.	+[a]	.	.
21 (Extension)	.	.	.	.	.	.	.	.	.	+[a]	.
22 (Techniques)	.	.	.	.	.	.	.	.	.	+[a]	.
23 (Terrain)	.	.	.	.	.	.	.	.	.	.	.
24 (Organization)	.	.	.	.	.	.	.	.	.	.	.
25 (School)	.	.	.	.	.	.	.	.	.	.	+[a]
26 (Enterprises)	.	.	.	.	.	.	.	.	.	.	+[a]
27 (Labor)	–	–	–	±[a]	±	–	–	–	–	–	±[a]
28 (Nutrition)	+	+	+	+	+	+	+	+	+	+	+
29 (Farmland)	.	.	.	.	.	.	.	.	.	.	.
30 (Inequity)	.	.	.	.	.	.	.	.	.	.	.
31 (Drug abuse)	+	+	+	±[a]	±	+	+	+	+	+	+

TABLE 4.5 (continued)
Impact of Gitangu Community's Goals Based on Pulse Process Analysis

Vertex		Community Goals										
		Pests and Diseases	Feed Quality	Roads	Employment	Soil Fertility	Water Availability	Human Health	Security	Market	Extension	Secondary Schools
Totals												
+	126	14	16	15	5	10	8	8	8	16	16	10
–	8	1	1	1	0	0	1	1	1	1	1	0

+ Positive impact; – negative impact; ± ambivalent; . no impact
[a] Impacts that are not sensitive to weight changes

4.3.3 Kiawamagira

Figure 4.5 is a cognitive map depicting relationships among factors influencing health and sustainability as perceived by residents of Kiawamagira village. Vertex 2 has cognitive centrality, with a total degree of 15, followed by vertices 1, 17, and 24, each with a total degree of 7. None of the vertices is a sink, but nine of them (3, 8, 15, 16, 20, 33, 34, 35, 37) are sources. Vertex 35 was ambivalent, being a source and having both positive (providing employment and manure) and negative (contributing to the pollution of the stream) impacts.

The digraph is balanced with reference to community goals, producing no indeterminate or ambivalent impacts (Table 4.6). The impacts of community goals increase to 107 if the arc [31, 30] is inverted. This also reduces the negative impacts to 0. Removing arcs [2, 29], [24, 2], and [28, 2] reduces the positive impacts of community goals to 79, 81, and 83, respectively, while reducing the negative impacts to 1, 8, and 8, respectively. Inverting the arcs [24, 2], [1, 2], [2, 5], and [12, 2] reduces the positive impacts of community goals to 75, 79, 81, and 82, respectively, while increasing the negative impacts to 23, 21, 19, and 16, respectively. It is unstable under all simple pulse processes if all arcs are given unit weight and time lag. The largest eigenvalue under this process is 2.58. Simple autonomous pulses, with equal weights and time lags on each arc, result in impacts similar to those determined through geometric analysis since the digraph is balanced. Because of this, no impacts are sensitive to changes in the weight.

There are two main strong components. The first consists of vertices 1, 2, 5, 25, 27, 28, 29, 30, and 31 interlinked into 7 two-arc and 2 three-arc positive-feedback loops. The second component comprises vertices 6, 7, and 24. Among the simplest stabilizing strategies for the first strong component is inverting any 3 two-arc cycles linked to vertex 2. The second strong component is pulse stable under all simple autonomous pulse processes. This component becomes value stable if arc [6, 24] or arc [24, 6] is removed.

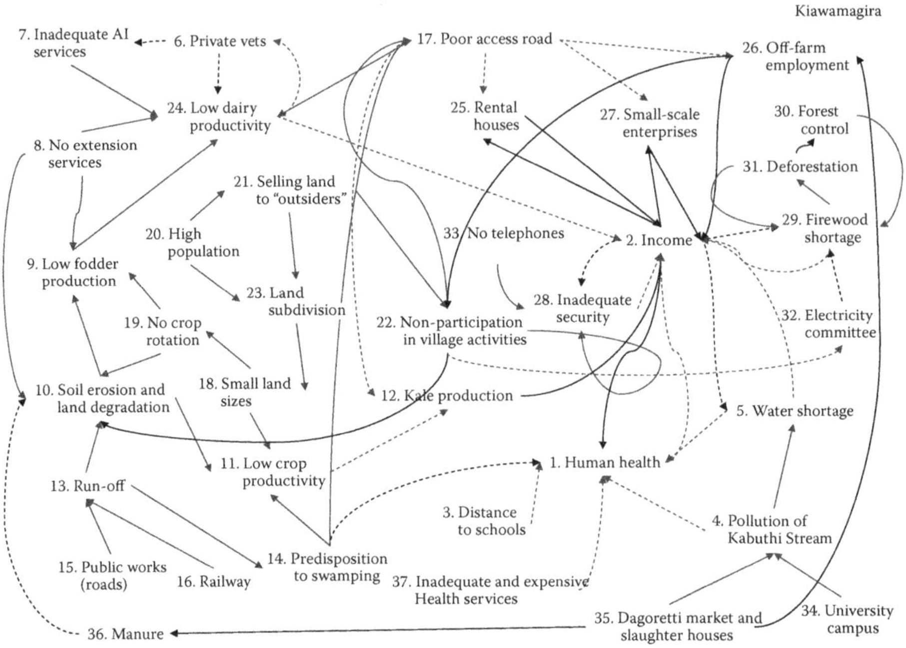

FIGURE 4.5 A cognitive map depicting perceptions of factors influencing health and sustainability in Kiawamagira intensive survey site, Kiambu Village, Kenya, 1997. See CD for color image. AI, artificial insemination.

4.3.4 Mahindi

The cognitive map depicting perceptions of the Mahindi community is shown in Figure 4.6. With a total degree of 12, vertex 4 has cognitive centrality. Other vertices with high total degrees are 33, 19, and 26, with total degrees of 7, 6, and 6, respectively. There are 12 sources (1, 3, 13, 15, 16, 17, 22, 23, 27, 32, 35, and 36) and 3 sinks (5, 30, and 37). Of the six key community goals, employment, water availability, and nursery schools show no impact on other vertices in the digraph (Table 4.7).

Water quality, health care, and roads have ambivalent impacts on vertices 7 and 8 owing to the perceived negative impacts of the arc [6, 7] and path imbalance between vertices 4 and 8. Inverting or removing arc [6, 8] increases the positive impacts of community goals to 40. Similar results are obtained with removal of arc [8, 7]. Removing arcs [9, 4], [18, 33], and [4, 18] reduces the positive impacts of community goals to 21, 26, and 26, respectively. Inverting arcs [9, 4], [18, 33], [4, 18], and [32, 19] reduces the positive impacts to 21, 22, 22, and 26, respectively, while increasing negative impacts to 17, 16, 16, and 4.

Assuming unit weight and time lags on each arc, the digraph is unstable under all simple autonomous pulse processes, with 1.25 the highest eigenvalue. The impacts of community goals based on these processes are shown in Table 4.8. Employment, water availability, and nursery schools have no impacts other than on their

TABLE 4.6
Impact of Kiawamagira Community's Goals Based on Geometric Analysis

	Community Goals									
Vertex	Distance to schools	Water availability	AI services	Crop productivity	Roads	Dairy productivity	Security	Fuel	Telephones	Healthcare
1 (Health)	+	+	+	+	+	+	+	+	+	+
2 (Income)	+	+	+	+	+	+	+	+	+	+
3 (Schools)	+	.	.	.	.	.	.	.	.	.
4 (Streams)	.	.	.	.	.	.	.	.	.	.
5 (Water)	+	+	+	+	+	+	+	+	+	+
6 (Vets)	.	.	+	.	+	+	.	.	.	.
7 (AI)	.	.	+	.	+	+	.	.	.	.
8 (Extension)	.	.	.	.	.	.	.	.	.	.
9 (Fodder)	.	.	.	.	.	.	.	.	.	.
10 (Soil)	.	.	.	.	.	.	.	.	.	.
11 (Crops)	.	.	.	+	.	.	.	.	.	.
12 (Kale)	.	.	.	+	+	.	.	.	.	.
13 (Runoff)	.	.	.	.	.	.	.	.	.	.
14 (Swamping)	.	.	.	.	.	.	.	.	.	.
15 (Public works)	.	.	.	.	.	.	.	.	.	.
16 (Railway)	.	.	.	.	.	.	.	.	.	.
17 (Roads)	.	.	.	.	+	.	.	.	.	.
18 (Land)	.	.	.	.	.	.	.	.	.	.
19 (Crop rotation)	.	.	.	.	.	.	.	.	.	.
20 (Population)	.	.	.	.	.	.	.	.	.	.
21 (Land sale)	.	.	.	.	.	.	.	.	.	.
22 (Participation)	.	.	.	.	.	.	.	.	.	.
23 (Subdivisions)	.	.	.	.	.	.	.	.	.	.

(continued on next page)

TABLE 4.6 (continued)
Impact of Kiawamagira Community's Goals Based on Geometric Analysis

	Community Goals									
Vertex	Distance to schools	Water availability	AI services	Crop productivity	Roads	Dairy productivity	Security	Fuel	Telephones	Healthcare
24 (Dairy)	.	.	+	.	+	+	.	.	.	.
25 (Rentals)	+	+	+	+	+	+	+	+	+	+
26 (Off-farm)	.	.	.	.	+	.	.	.	.	.
27 (Enterprises)	+	+	+	+	+	+	+	+	+	+
28 (Security)	+	+	+	+	+	+	+	+	+	+
29 (Fuel)	+	+	+	+	+	+	+	+	+	+
30 (Forest)	–	–	–	–	–	–	–	–	–	–
31 (Deforestations)	+	+	+	+	+	+	+	+	+	+
32 (Electricity)	.	.	.	.	.	.	.	.	.	.
33 (Telephones)	.	.	.	.	.	.	.	.	+	.
34 (University)	.	.	.	.	.	.	.	.	.	.
35 (Abattoirs)	.	.	.	.	.	.	.	.	.	.
36 (Manure)	.	.	.	.	.	.	.	.	.	.
37 (Healthcare)	.	.	.	.	.	.	.	.	.	+
Totals										
+ 97	9	8	11	10	14	11	8	8	9	9
- 10	1	1	1	1	1	1	1	1	1	1
Problem ranking	?[a]	2	6	3?[b]	1	3?[b]	5	4	8	7
Goal status	0	0	0	1	1	1	2	0	0	1

Key: + Positive impact; – Negative impact; ± Ambivalent; . No impact; # Indeterminate
[a] Not ranked/identified in the initial workshop
[b] Problem stated as low farm productivity during initial workshop

corresponding vertices. The impacts of the rest of the goals are predominantly ambivalent, with roads having negative impacts on vertices 29 and 30 and positive impacts on vertices 5, 19, 20, 31, and 32. Of the 48 impacts of community goals, 21 are not sensitive to increases in weights, but only 7 of them are nonambivalent and

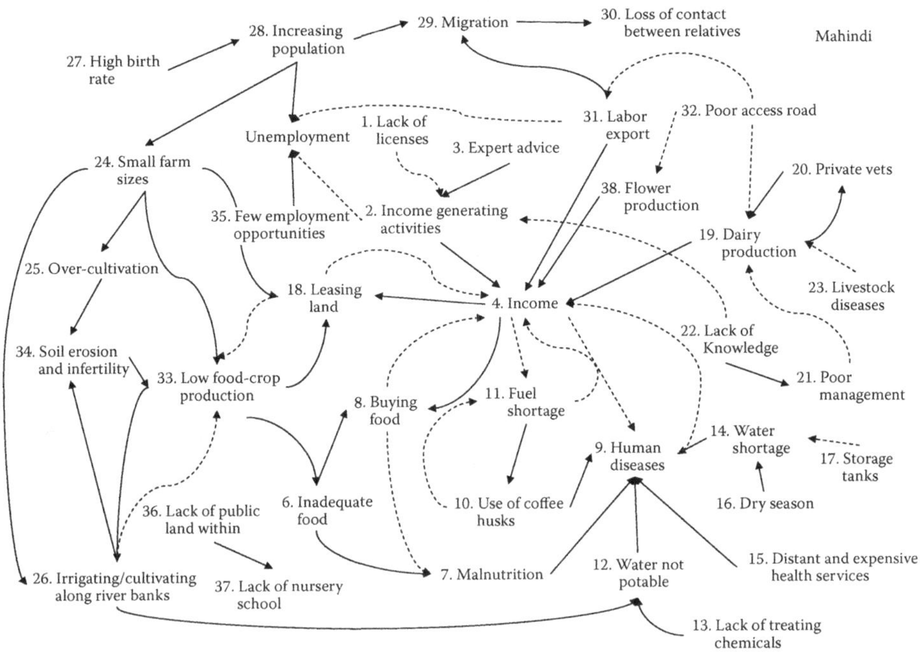

FIGURE 4.6 A cognitive map depicting perceptions of factors influencing the health and sustainability in Mahindi intensive survey site, Kiambu District, Kenya, 1997. See CD for color image and key.

indirect (Table 4.8). The impacts are most sensitive to increases in the weights of any of the arcs in the two-arc cycles linking vertex 4 to vertices 11 and 9.

The digraph consists of two main strong components. The first has 12 vertices (4, 6, 7, 8, 9, 10, 11, 12, 18, 20, 33, and 34) in six negative- and eight positive-feedback loops. The second has two (19 and 20) vertices in a positive-feedback loop. The first strong component is pulse stable. The simple value-stabilizing strategies for this component include removing arc [18, 33] and then either arc [20, 33] or [34, 33]. The second component is pulse stable as well and can be value stabilized by inverting any one of the two arcs.

4.3.5 Gikabu

The cognitive map produced by participants from Gikabu is shown in Figure 4.7. Vertex 9 has cognitive centrality, with a total degree of 14, followed by vertices 26, 29, and 18, with total degrees of 7, 7, and 6, respectively. Nine of the vertices (1, 10, 16, 19, 20, 21, 22, 32, 33) are sources, but there are no sinks. Table 4.9 shows the impacts of community goals in Gikabu village based on a geometric analysis of the digraph. Tea markets had indeterminate effects on most other vertices due to the presence of 2 three-arc paths from vertex 3 to 9. It had negative impacts on vertex 5 and 6. Inverting or removing arc [6, 7] increases the positive impacts of community goals to 130, while removing arc [7, 9] increases the impacts to 129. Inverting any one of arcs [15, 9], [9, 26], [26, 29], and [13, 14] reduces the positive impacts of

TABLE 4.7
Impact of Mahindi Community's Goals Based on Geometric Analysis

	Community Goals					
Vertex	Employment	Water Quality	Water Availability	Health Care	Roads	Nursery School
1 (Licenses)	.	.	.	.	.	.
2 (Activities)	.	.	.	.	.	.
3 (Experts)	.	.	.	.	.	.
4 (Income)	.	+	.	+	+	.
5 (Employment)	+	.	.	.	+	.
6 (Food)	.	+	.	+	+	.
7 (Nutrition)	.	±	.	±	±	.
8 (Buying food)	.	±	.	±	±	.
9 (Diseases)	.	+	.	+	±	.
10 (Husks)	.	+	.	+	+	.
11 (Fuel)	.	+	.	+	+	.
12 (Water quality)	.	+	.	+	+	.
13 (Chemicals)	.	.	.	.	.	.
14 (Water)	.	.	+	.	.	.
15 (Health care)	.	.	.	+	.	.
16 (Season)	.	.	.	.	.	.
17 (Tanks)	.	.	.	.	.	.
18 (Leasing)	.	+	.	+	+	.
19 (Dairy)	.	.	.	.	+	.
20 (Vets)	.	.	.	.	+	.
21 (Management)	.	.	.	.	.	.
22 (Knowledge)	.	.	.	.	.	.
23 (Livestock diseases)	.	.	.	.	.	.
24 (Farm size)	.	.	.	.	.	.
25 (Cultivation)	.	.	.	.	.	.
26 (Riverbanks)	.	+	.	+	+	.
27 (Birth rate)	.	.	.	.	.	.
28 (Population)	.	.	.	.	.	.
29 (Migration)	.	.	.	.	–	.
30 (Contact)	.	.	.	.	–	.
31 (Labor)	.	.	.	.	+	.
32 (Roads)	.	.	.	.	+	.
33 (Crops)	.	+	.	+	+	.
34 (Soil)	.	+	.	+	+	.
35 (Jobs)	.	.	.	.	.	.

TABLE 4.7 (continued)
Impact of Mahindi Community's Goals Based on Geometric Analysis

Vertex		Community Goals					
		Employment	Water Quality	Water Availability	Health Care	Roads	Nursery School
36 (Public land)		.	.	.	.	.	.
37 (School)		.	.	.	.	.	+
38 (Flowers)		.	.	.	.	+	.
Total							
+	36	1	9	1	10	14	1
–	2	0	0	0	0	2	0
Problem ranking		2	?	3	4	1	5
Goal status		0	0	0	0	1	0

+ Positive impact; – negative impact; ± ambivalent; . no impact; # indeterminate

TABLE 4.8
Impact of Mahindi Community's Goals Based on Pulse Analysis

Vertex	Community Goals					
	Employment	Water Quality	Water Availability	Health Care	Roads	Nursery School
1 (Licenses)	.	.	.	.	.	.
2 (Activities)	.	.	.	.	.	.
3 (Experts)	.	.	.	.	.	.
4 (Income)	.	±	.	±	±	.
5 (Employment)	+[a]	.	.	.	+[a]	.
6 (Food)	.	±	.	±	±	.
7 (Nutrition)	.	±[a]	.	±[a]	±[a]	.
8 (Buying food)	.	±[a]	.	±[a]	±[a]	.
9 (Diseases)	.	±	.	±	±	.
10 (Husks)	.	±	.	±	±	.
11 (Fuel)	.	±	.	±	±	.
12 (Water quality)	.	±	.	±	±	.
13 (Chemicals)	.	.	.	.	.	.
14 (Water)	.	.	+[a]	.	.	.
15 (Health care)	.	.	.	+[a]	.	.

(continued on next page)

TABLE 4.8 (continued)
Impact of Mahindi Community's Goals Based on Pulse Analysis

	Community Goals					
Vertex	**Employment**	**Water Quality**	**Water Availability**	**Health Care**	**Roads**	**Nursery School**
16 (Season)	.	.	.	.	.	.
17 (Tanks)	.	.	.	.	.	.
18 (Leasing)	.	±[a]	.	±[a]	±[a]	.
19 (Dairy)	.	.	.	.	+[a]	.
20 (Vets)	.	.	.	.	+[a]	.
21 (Management)	.	.	.	.	.	.
22 (Knowledge)	.	.	.	.	.	.
23 (Livestock diseases)	.	.	.	.	.	.
24 (Farm size)	.	.	.	.	.	.
25 (Cultivation)	.	.	.	.	.	.
26 (Riverbanks)	.	±	.	±	±	.
27 (Birth rate)	.	.	.	.	.	.
28 (Population)	.	.	.	.	.	.
29 (Migration)	.	.	.	.	–[a]	.
30 (Contact)	.	.	.	.	–[a]	.
31 (Labor)	.	.	.	.	+[a]	.
32 (Roads)	.	.	.	.	+[a]	.
33 (Crops)	.	±	.	±	±	.
34 (Soil)	.	±	.	±	±	.
35 (Jobs)	.	.	.	.	.	.
36 (Public land)	.	.	.	.	.	.
37 (School)	.	.	.	.	.	+[a]
38 (Flowers)	.	.	.	.	+[a]	.
Totals						
+ 10	1	0	1	1	6	1
– 2	0	0	0	0	2	0

+ Positive impact; – negative impact; ± ambivalent; . no impact

[a] Impacts not sensitive to change in weights

community goals to 59, 69, 72, and 75, respectively, while increasing the negative impacts to 24, 41, 33, and 11, respectively. Removing any one of arcs [9, 26], [26, 29], [12, 13], [31, 9], and [15, 9] reduces the positive impacts of community goals to 79, 87, 94, 97, and 96, respectively.

The digraph is unstable under pulse processes if unit weight and time lag are assumed for each of the arcs, with 2.67 the largest eigenvalue. Under the pulse

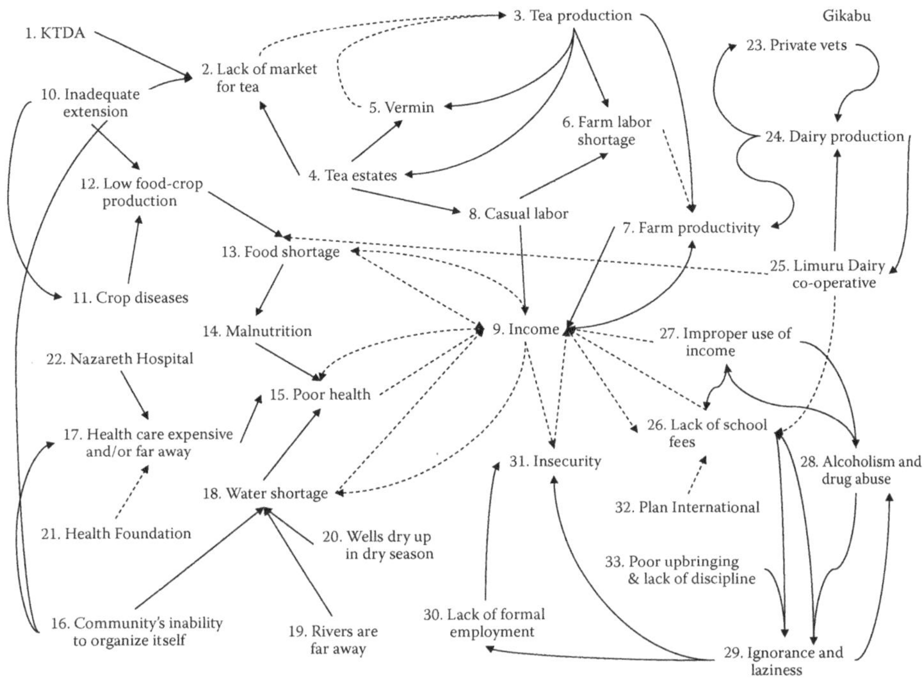

FIGURE 4.7 A cognitive map depicting perceptions of factors influencing health and sustainability in Gikabu intensive survey site, Kiambu District, Kenya, 1997. (KTDA: Kenya Tea Development Authority.) See CD for color image and key.

process, the impacts of all community goals except tea markets remain as shown in Table 4.10. Tea market produces oscillating impacts at most vertices that are indeterminate through the geometric analysis except vertices 15 and 31, where it has positive impacts. The only impacts that are sensitive to changes in the weight of the arcs are those of tea market on vertices 9, 13, 14, 15, 18, and 26 through 31.

The digraph consists of three main strong components. The first is pulse stable and consists of 4 vertices (2, 3, 4, and 5) in two negative-feedback loops. The second, consisting of 12 vertices (7, 9, 13, 14, 15, 18, 26, 27, 28, 29, 30, and 31) in 15 positive-feedback loops, is unstable. The third component comprises vertices 23, 24, and 25 in two positive-feedback loops and is pulse stable. The first strong component can be value stabilized by removing any one of its arcs except [4, 5]. Among the simple ways of value stabilizing the second component is removal of arcs [9, 26] and [9, 18], inversion of arc [9, 13], followed by inversion of any one of arcs [27, 28], [26, 29], [28, 29], and [29, 28]. The third component can be value stabilized by inverting any one of its arcs.

4.3.6 Thiririka

In the cognitive map depicting perceptions of the Thiririka community on factors influencing health and sustainability (Figure 4.8), vertex 7 has cognitive centrality with a total degree of 10. Other vertices with high total degrees are 26, 3, and 4, with

TABLE 4.9
Impact of Gikabu Community's Goals Based on Geometric Analysis

	Community Goals									
Vertex	Tea Market	Extension	Crop Diseases	Food Shortage	Health	Health Care	Water	Drugs	Employment	Security
1 (KTDA)	.	.	.	.	.	.	.	.	.	.
2 (Tea market)	+	.	.	.	.	.	.	.	.	.
3 (Tea)	+	.	.	.	.	.	.	.	.	.
4 (Tea estates)	+	.	.	.	.	.	.	.	.	.
5 (Vermin)	–	.	.	.	.	.	.	.	.	.
6 (Labor shortage)	–	.	.	.	.	.	.	.	.	.
7 (Productivity)	#	+	+	+	+	+	+	+	+	+
8 (Farm labor)	+	.	.	.	.	.	.	.	.	.
9 (Income)	#	+	+	+	+	+	+	+	+	+
10 (Extension)	.	+	.	.	.	.	.	.	.	.
11 (Crop diseases)	.	+	+	.	.	.	.	.	.	.
12 (Crops)	.	+	+	.	.	.	.	.	.	.
13 (Food)	#	+	+	+	+	+	+	+	+	+
14 (Nutrition)	#	+	+	+	+	+	+	+	+	+
15 (Health)	#	+	+	+	+	+	+	+	+	+
16 (Organization)	.	.	.	.	.	.	.	.	.	.
17 (Health care)	.	.	.	.	.	+	.	.	.	.
18 (Water)	#	+	+	+	+	+	+	+	+	+
19 (Rivers)	.	.	.	.	.	.	.	.	.	.
20 (Wells)	.	.	.	.	.	.	.	.	.	.
21 (Foundation)	.	.	.	.	.	.	.	.	.	.
22 (Hospital)	.	.	.	.	.	.	.	.	.	.
23 (Vets)	.	.	.	.	.	.	.	.	.	.
24 (Dairy)	.	.	.	.	.	.	.	.	.	.
25 (Co-op)	.	.	.	.	.	.	.	.	.	.
26 (Fees)	#	+	+	+	+	+	+	+	+	+
27 (Income use)	#	+	+	+	+	+	+	+	+	+
28 (Drug abuse)	#	+	+	+	+	+	+	+	+	+
29 (Ignorance)	#	+	+	+	+	+	+	+	+	+
30 (Unemployment)	#	+	+	+	+	+	+	+	+	+
31 (Security)	#	+	+	+	+	+	+	+	+	+
32 (Plan)	.	.	.	.	.	.	.	.	.	.
33 (Discipline)	.	.	.	.	.	.	.	.	.	.

TABLE 4.9 (continued)
Impact of Gikabu Community's Goals Based on Geometric Analysis

Vertex		Community Goals									
		Tea Market	Extension	Crop Diseases	Food Shortage	Health	Health Care	Water	Drugs	Employment	Security
Totals											
+	118	4	15	14	12	12	13	12	12	12	12
–	2	2	0	0	0	0	0	0	0	0	0
Problem ranking		6	7	5	9	3	3	1	8	4	2
Goal status		2	2	1	1	0	0	0	0	0	3

+ Positive impact; – negative impact; ± ambivalent; . no impact
KTDA: Kenya Tea Development Authority

TABLE 4.10
Impacts of Thiririka Community's Goals Based on Geometric Analysis

Vertex	Community Goals										
	Human Health	Security	Agrochemicals	Crop Diseases	Health Care	Seeds Quality	Soil Productivity	Community Organization	Leadership	Land Grabbing	Extension
1 (Water logging)	.	.	.	.	.	.	.	.	.	.	.
2 (Water quality)	.	.	.	.	.	.	.	.	.	.	.
3 (Income)	+	+	±	+	+	+	+	+	+	+	+
4 (Health)	+	+	±	+	+	+	+	+	+	+	+
5 (Poverty)	+	+	±	+	+	+	+	+	+	+	+
6 (Security)	+	+	±	+	+	+	+	+	+	+	+
7 (Horticulture)	.	.	–	+	.	+	+	+	+	+	+
8 (Water)	.	.	.	.	.	.	.	.	.	.	.
9 (Productivity)	+	+	#	+	+	+	+	+	+	+	+
10 (Chemicals)	.	.	+	+	.	.	.	.	.	.	.
11(Crop diseases)	.	.	–	+	.	.	.	.	.	.	.
12 (Health care)	.	.	.	.	+	.	.	.	.	.	.
13 (Fodder)	.	.	.	.	.	.	.	.	.	.	.
14 (Dairy)	.	.	.	.	.	.	.	+	+	+	+

(continued on next page)

TABLE 4.10 (continued)
Impacts of Thiririka Community's Goals Based on Geometric Analysis

	Community Goals										
Vertex	Human Health	Security	Agrochemicals	Crop Diseases	Health Care	Seeds Quality	Soil Productivity	Community Organization	Leadership	Land Grabbing	Extension
15 (Employment)	.	.	–	+	.	+	+	+	+	+	+
16 (Firewood)	.	.	.	.	.	.	.	.	.	.	.
17 (Forest)	.	.	.	.	.	.	.	.	.	.	.
18 (Logging)	.	.	.	.	.	.	.	.	.	.	.
19 (Co-op)	.	.	.	.	.	.	.	+	+	+	+
20 (Seed)	.	.	.	.	.	+	.	.	.	.	.
21 (Soil)	.	.	.	.	.	.	+	.	.	.	.
22 (Frost)	.	.	.	.	.	.	.	.	.	.	.
23 (Unity)	.	.	.	.	.	.	.	+	+	+	+
24 (Leadership)	.	.	.	.	.	.	.	+	+	+	+
25 (Grabbing)	.	.	.	.	.	.	.	+	+	+	+
26 (Knowledge)	.	.	.	.	.	.	.	+	+	+	+
27 (Relationships)	.	.	.	.	.	.	.	.	.	.	.
28 (Market)	.	.	.	.	.	.	.	+	+	+	+
29 (History)	.	.	.	.	.	.	.	.	.	.	.
30 (Extension)	.	.	.	.	.	.	.	.	.	.	+
31 (Go-downs)	.	.	.	.	.	.	.	.	.	.	.
Totals											
+ 99	5	5	1	9	6	8	8	14	14	14	15
– 3	0	0	3	0	0	0	0	0	0	0	0
Problem ranking	2	1	8	10	2	4	9	5	7	3	6
Goal status	1	3	1	1	0	0	1	1	1	2	2

+ Positive impact; – negative impact; ± ambivalent; . no impact; # indeterminate

total degrees of 8, 7, and 7, respectively. Nine of the vertices (1, 8, 12, 20, 21, 22, 29, 30, and 31) are sources, and none is a sink.

The impacts of community goals based on a geometric analysis are shown in Table 4.10. Reduction in agrochemical usage results in negative impacts on vertices 7 and 11; ambivalent impacts on vertices 3, 4, 5, and 6; and indeterminate impacts on vertex 9. The ambivalent and indeterminate impacts of agrochemical use result from path imbalance from vertex 10 to vertex 9. The positive impacts of community goals increase to 107 when the sign of arc [10, 11] is inverted and to 104 if it is removed.

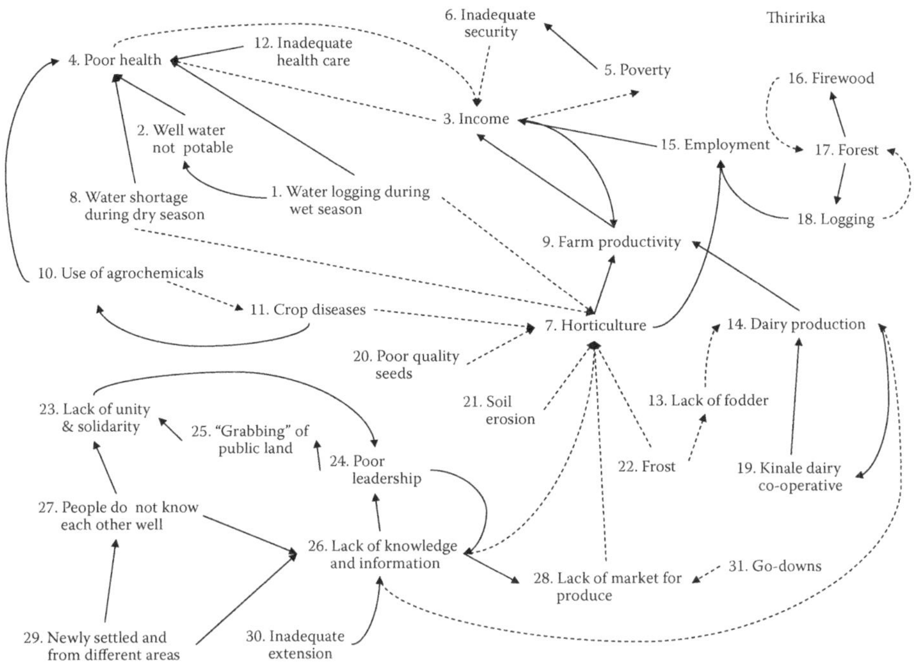

FIGURE 4.8 A cognitive map depicting perceptions of factors influencing the health and sustainability in Thiririka intensive survey site, Kiambu District, Kenya, 1997. See CD for color image and key.

Positive impacts also increase (to 102) if arc [11, 7] is removed. In all three cases, the negative impacts are eliminated. Inversion of any one of the arcs [7, 15], [15, 3], and [7, 9] reduces the positive impacts of community goals to 58, 64, and 64, respectively, while the negative impacts are unaffected except for arc [7, 15], for which the negative impacts increase to 9.

The digraph is unstable under all simple autonomous pulse processes assuming unit weight and time lags on all arcs of the digraph. The largest eigenvalue based on unit arc weights and time lags is 1.62. Based on this pulse process, agrochemical use and crop diseases have oscillating impacts on vertices 7, 9, 10, and 11. Crop diseases also have oscillating impacts on vertex 15. Impacts of community goals on other vertices remain as in Table 4.10. Most (97/107) of the impacts are not sensitive to increases in weights. The 10 that are weight dependent are the impacts of agrochemical use and crop diseases on vertices 3, 4, 5, 6, and 8. Impacts are most sensitive to increases in the weights of arcs [11, 10] and [10, 11].

The digraph consists of five strong components, two of which have two-arc feedback loops involving vertices 10 and 11 in one (negative) and 14 and 19 in the other (positive). The largest strong component, comprising vertices 3, 4, 5, 6, and 9 in 2 two-arc and on three-arc positive-feedback loops, is unstable. One of the simplest strategies for value stabilizing this component is inverting any one of the arcs in any of the two-arc feedback loops. The next largest component comprises vertices 23,

24, 25, and 26 in two positive-feedback loops and is pulse stable. This component can be value stabilized by inverting any one of arcs [1, 2], [2, 4], and [3, 1]. The fourth component is value stable and comprises vertices 16, 17, and 18 in 2 two-arc negative-feedback loops.

4.4 DISCUSSION

4.4.1 Construction of Cognitive Maps

The idea of cognitive maps was easily understood and utilized by communities. This may be a reflection of the fact that the maps are a much easier way of depicting their perceptions, which in turn indicates that communities are aware of the high degree of interrelationships among factors that influence the health and sustainability of their agroecosystems. The use of a metaphor as a guide in the selection of concepts to be included in the map was successful in all the communities.

Dividing the participants into smaller groups during construction of the cognitive maps allowed for the expression of different perspectives and the active involvement of most participants. The visual nature of the cognitive maps makes it easy to engender debate, thus providing a basis for debate and consensus building among the participants and the creation of a synthesis cognitive map. However, the end product is a compromise between the various views and does not necessarily capture all the divergent perspectives. In this study, relationships and concepts that were not unanimously agreed on by all participants were left out of the final cognitive map. Ideally, cognitive maps representing the most divergent of perspectives should also be analyzed and the conclusions compared and offered for debate by communities.

Cognitive maps were largely in agreement with the findings of the initial village workshops. This was remarkable given that the cognitive maps were drawn several months after the initial workshops. Important discrepancies, however, were present in some cognitive maps, especially with regard to the role of some institutions and in the number and descriptions of some of the problems. In Mahindi village, water quality was added as a problem in the cognitive maps. In Kiawamagira, distance to schools was added, while crop and dairy productivity were mentioned as separate issues. These changes were most likely due to a reevaluation of the problem situation rather than an inaccuracy in the findings of the village workshops.

4.4.2 Use of Signed Digraphs

The analysis of the communities' cognitive maps using graph theory was constrained by two key limitations. The first was that it was difficult to assign weights to relationships between many of the concepts used in the cognitive maps. Even where this was possible, communities found it difficult to apply mathematical ideas to concepts and relationships that they perceived mostly in qualitative terms. The other constraint was that it was difficult to state concepts and relationships in such a way that all the arcs in the digraph reflect equal time lags. In this study, participants were made aware of the need to state relationships in a way that makes the arcs have more or less equal time lags, but many participants were unable or unwilling to put this constraint in their maps. The digraphs are therefore most useful for short-to-medium-term analysis that can be updated iteratively. Using geometric analytical techniques as well as sensitivity analysis, useful insights can still be obtained from the less-detailed digraphs derived from these cognitive maps. However, the conclusions from these analyses are less detailed than would be with complete and detailed digraphs. In an action research process, cognitive maps can be reevaluated and updated in each action research cycle as the local theory develops. In this way, more details—both structural and numerical—can be added to enable detailed analyses.

In this study, an additional quality—impact—of the relationships between vertices was included. The rationale for this was that the purpose of community goals and objectives was to minimize the negative impact of problem situations. Because of the interrelationships among factors, minimizing negative impact involves trade-offs because some of the solutions may in turn generate negative impacts. The goal-seeking behavior of the system can therefore be seen as maximizing the positive impact of community goals through changes in structure and the application of pulses to the system. In addition, the digraph assumes that the system dynamics are linear. Although this is not the case for most of the relationships, the linear model is likely to be a suitable generalization of the processes over the short term.

4.4.3 Geometric Analyses

Income, farm productivity, and human health were consistently among the factors that had high cognitive centrality. This would be expected for agricultural communities in a largely subsistence farming system. The inclusion of labor as an important system output and its export as a coping strategy, however, may be an indication that subsistence was heavily constrained. This conclusion was further augmented by the existence of several sources in all the digraphs, many of which related to infrastructure and other cash-economy-dependent components. While some of these may be due to cognitive imbalances, they also indicate (largely) a perceived dependency on external intervention. A good example was in Mahindi village, where use of water tanks was seen as ameliorating water shortage. An agent external to the community had donated these tanks to some households. No connection was made between income and availability of tanks. Surprisingly, although many of the sources were recognized as public goods that should be provided through a taxation system, participants in all communities were unanimous that those connections do not exist in reality.

In nearly all communities, the rank of community goals based on their total impact was different from the ranking during the initial workshops, suggesting that either the geometric process was not confluent with the community's cognitive processes or the impact of goals was not the predominant criterion for ranking. Unexpectedly, the ranking of community goals based on their total degrees was remarkably confluent with that in the initial village workshops, suggesting that the complexity of interrelationships may have been an important ranking criterion. It is therefore not surprising that all communities reconsidered their ranking after drawing the cognitive maps. It would have been useful to provide the communities with the results of the geometric analysis during this reevaluation. Unfortunately, this was not possible in this study.

Geometric analysis was also useful in analyzing the predicates on which the community goals were based. An arc that, on removal or inversion, results in increase in the number of positive impacts of community goals can be interpreted as representing either a constraint or a coping strategy. For example, the arc [6, 9] in the cognitive digraph from Githima represents negative consequences of the use of agrochemicals on human health, which constrain their use as a means to increase farm productivity. In contrast, the arc [12, 27] in the Gitangu digraph represents the trend for younger people to seek formal employment outside the village as a result of reduced availability of farmland and therefore represents a coping strategy. Arcs with removal that results in the reduction of the positive impacts of community goals can be interpreted as representing the relationships on which the community goals are based. An example is Kiawamagira village, where removal of arc [2, 29], analogous to the assertion that improving dairy productivity would have no effect on incomes, severely reduces the positive impacts of community goals. These findings therefore provide an objective and reproducible approach to assessing agroecosystem health and sustainability goals and objectives and the relationships between them.

4.4.4 Pulse Process Models

Analyzing the impact of community goals using geometric analyses is limited by the existence of indeterminacies resulting from some kinds of path imbalances. The pulse process provides some indication of what the impacts would be under certain conditions. In this study, the usefulness of this approach was constrained by the inability to obtain reasonably accurate weights for the arcs in the digraphs. In addition, the complexity of the digraphs makes it difficult to assess sensitivity to all possible weight structures. However, the assumption of unit weights and time lags for the arcs may still provide some useful insights. Also, understanding the sensitivity of the impact of community goals on the changes in weights of a particular arc provides a means of generating hypotheses regarding which relationships are likely to be relatively more important regarding the system's health and sustainability. In the Mahindi cognitive map, for example, increases in the weights of any of the two-arc cycles linking vertex 4 to vertices 11 and 9 stabilizes many of the oscillating impacts of community goals, with the direction of stability depending on the sign and weight structure of the arcs.

An interesting feature was that the reranking of community goals that was carried out following the cognitive map exercise was confluent with the summary of impacts based on the pulse process model. This probably indicates that the communities perceived the relationships to be more or less linear and the arcs as bearing unit weight and time lags. It would be gainful to provide the results of the current analysis to the communities for discussion and to give them an opportunity to modify the structure of the cognitive maps or the ranking of their goals based on these findings.

4.4.5 Assessment of Value Stability

Stability was interpreted based on the assumption that there are always some limits to growth in most real-world situations (Perry, 1983). This limit manifests itself—in most cases—as value stability. Absence of value stability (which implies pulse stability) or at the very least quasi stability can therefore be interpreted as a reflection of cognitive imbalance (failure to consider the opposite effects of a relationship) and hence an inaccuracy in the cognitive map as representation of the domain modeled. Cognitive maps dealing with goals and objectives are likely to have cognitive imbalances due to the tendency to perceive regenerative or degenerative spirals related to goals or problems. A regenerative spiral is when no limits to goals and objectives are foreseen, while a degenerative spiral is when no limits to deterioration of the problem situation can be foreseen.

Understanding the ways in which stability can be imparted to the digraph may provide insights on the relationships on which these imbalances are predicated. As an example, the Gikabu digraph becomes value stable if arcs [2, 3], [9, 26], and [9, 18] are removed while arcs [25, 24], [9, 13], and [26, 29] are inverted. This indicates that the regenerative spiral in this digraph is predicated on the perception that improvements in the markets would produce unbounded and direct increases in tea production and the same for increased efficiency of the dairy cooperative on dairy production, among others. The different stabilizing strategies represent the possible scenarios in which the perceived spirals do not exist. The impact of community goals based on the digraph resulting from these stabilizing strategies can be used to assess community goals in the absence of cognitive imbalances.

REFERENCES

Ackoff, R.L. (1980). The systems revolution. In *Organizations as Systems*. Lockett, M., and Spear, R., eds. London: Open University Press, pp. 26–33.

Axelrod, R. (1976a). The analysis of cognitive maps. In *Structure of Decision: The Cognitive Maps of Political Elites*. Axelrod, R., ed. Princeton, NJ: Princeton University Press, pp. 55–73.

Axelrod, R. (1976b). The cognitive mapping approach to decision making. In *Structure of Decision: The Cognitive Maps of Political Elites*. Axelrod, R., ed. Princeton, NJ: Princeton University Press, pp. 3–17.

Bang-Jensen, J., and Gutin, G. (2001). *Digraphs: Theory, Algorithms and Applications*. Springer-Verlag, New York. 754 pp.

Caws, P. (1988). *Structuralism: The Art of the Intelligible*. International Humanities Press, Atlantic Highlands, NJ, p. 276.

Dalkey, N.C. (1969). *The Delphi Method: An Experimental Study of Group Opinion*. RM-5888-PR. Rand, Santa Monica, CA.

Eden, C., Jones, S., and Sims, D. (1983). *Messing About in Problems: An Informal Structured Approach to their Identification and Management*. Pergamon Press, Oxford, U.K.

Flood, R.L., and Carson, E.R. (1993). *Dealing with Complexity: An Introduction to the Theory and Application of Systems Science*. New York: Plenum Press.

Harary, F., Norman, R.Z., and Cartwright, D. (1965). *Structural Models: An Introduction to the Theory of Directed Graphs*. Wiley, Chichester, U.K., p. 415.

Ison, R.L., Maiteny, P.T., and Carr, S. (1997). Systems methodologies for sustainable natural resources research and development. *Agricultural Systems* 55: 257–272.

Jeffries, C. (1974). Qualitative stability and digraphs in model ecosystems. *Ecology* 55: 1415–1419.

Klee, V. (1989). Sign-pattern and stability. In *Applications of Combinatorics and Graph Theory to the Biological and Social Sciences*. Roberts, F., ed. New York: Springer-Verlag, pp. 202–219.

Nozicka, J.G., Bonham, G.M., and Shapiro, M. J. (1976). Simulation techniques. In *Structure of Decision: The Cognitive Maps of Political Elites*. Axelrod, R., ed. Princeton, NJ: Princeton University Press, pp. 349–359.

Perry, E.L. (1983). A pulse process model of athletic financing. In *Modules in Applied Mathematics*. Lucas, W.F., Roberts, F.S., and Thrall, R.M., eds. New York: Springer-Verlag, pp. 141–164.

Puccia, C.J., and Levins, R. (1985). *Qualitative Modeling of Complex Systems*. Harvard University Press, Cambridge, MA. 259 pp.

Ridgley, M., and Lumpkin, C.A. (2000). The bi-polar resource-allocation problem under uncertainty and conflict: A general methodology for the public decision-maker. *Journal of Environmental Management* 59: 89–105.

Roberts, F.S. (1971). Signed digraphs and the growing demand for energy. *Environment and Planning* 5: 395–410.

Roberts, F.S. (1976a). *Discrete Mathematical Models, with Applications to Social, Biological, and Environmental Problems*. Prentice-Hall, Englewood Cliffs, NJ, p. 559.

Roberts, F.S. (1976b). Strategy for the energy crisis: The case of commuter transportation policy. In *Structure of Decision: The Cognitive Maps of Political Elites*. Axelrod, R., ed. Princeton, NJ: Princeton University Press, pp. 142–179.

Roberts, F.S., and Brown, T.A. (1975). Signed digraphs and the energy crisis. *American Mathematical Monthly* 82: 577–594.

Sorensen, A. (1978). Mathematical models in sociology. *Annual Review of Sociology* 4: 345–371.

5 Soft System Methodology in the Management of Agroecosystem Health and Sustainability Concerns of a Tropical Highlands Agroecosystem

5.1 INTRODUCTION

Hard system methods (also known as first-order cybernetics) are concerned with problem situations in which there is correspondence between the holon and things or phenomena in the real world. Soft system methods (second-order cybernetics), on the other hand, deal with situations in which such correspondence may not exist, with the holons articulated solely for the purpose of understanding and as shorthand for the mental framework of an individual with a unique experiential or cognitive history on the relevant real-world situation (Ison et al., 1997).

The objectives of a soft system analysis in agroecosystem health (AESH) and sustainability assessment are threefold. The first is to reveal the different, and sometimes conflicting, perspectives of stakeholders and the rationale behind each perception of a problem situation. This prepares the ground for mutual understanding and negotiation among the stakeholders, a prerequisite to any sustainable improvements. The second objective is to serve as the basis for evaluating potential management options. As the agroecosystem evolves over time, new aspects of the messy problem situation emerge, requiring new analysis and synthesis as well as rethinking the management options. The third objective is the evaluation of agroecosystem performance. This can be done by comparing and contrasting holon characteristics over time and space.

Several soft system approaches have been proposed (Flood and Carson, 1993), but the best documented is soft system methodology (SSM). SSM is a set of organized principles—based on systems thinking—that guide action in trying to manage messy problem situations. SSM follows two interacting modes of inquiry that together lead to the implementation of changes to improve the situation. One of these, the cultural stream, consists of three examinations of the problem situation (Checkland and Scholes, 1990). The first examines the intervention itself. The second examines the situation as a social system; the third examines it as a political system. The basic step in the second (logic-based) mode of inquiry is to formulate models, which it is hoped will be relevant to the real-world situation (Checkland and Scholes, 1990). The models are then compared with various perceptions of the real world, thus initiating debates and a process of negotiations and trade-offs that lead to purposeful actions aimed at improving the problem situation under scrutiny.

SSM uses particular kinds of holons—referred to as *human activity systems*—to model the problem situation. A human activity system is a set of named activities connected to make a purposeful whole (Checkland and Scholes, 1990). The holons are conceived as holistic ideal types of certain aspects of the problem situation rather than as accounts of it; it is taken for granted that no objective and complete account of a messy problem situation can be given (Bulow, 1989). Two kinds of human activity systems can be made: the primary task system and the issue-based system (Checkland and Wilson, 1980). Primary task systems are ones in which the elements and relationships map on to real-world institutionalized arrangements. Issue-based systems, on the other hand, are relevant to mental processes that are not embodied in formalized real-world arrangements. The distinction between primary task and issue-based systems is not absolute but rather more of opposite ends of a spectrum. The choice of a human activity system to represent a problem situation is always subjective (Checkland and Scholes, 1990), with the final choice depending on which model is deemed most relevant to the situation after the logical implications of all the choices have been evaluated.

A human activity system is built based on its root definition. A *root definition* expresses the core purpose of an activity system. That core purpose is always expressed as a transformation process in which some entity, the "input," is changed, or transformed, into some new form of that same entity, the "output." The transformation occurs because a purposeful action (or actions) A is (or are) taken on that entity. Such an action, being purposeful, will be an expression of the intention of some person or persons B. Since A is a human action, there will be someone C who takes the action. The action will have an impact on some person or group D, and it will be taking place in an environment E, which may place constraints on it. Since human autonomy is rarely total, there may be a person or group F who could stop the action being taken. In real life, the same person or persons could be one or more of the elements in B, C, D, or F since these represent roles and not individuals or groups playing them. The transformation and the actions taken are meaningful and rational given a particular perspective or worldview. A complete root definition of a human activity system therefore identifies the customer (D), the actor (C), the transformation, the worldview, the owner (B), and the environment of a particular activity system.

Several human activity systems can be built to represent different perspectives of a given world situation. In addition, activity systems can be built for intended purposeful actions, with several such systems representing the different perspectives that may exist among all the stakeholders. These models then serve as the basis for negotiation and consensus building as well as a guide to action, monitoring, and evaluation. This chapter describes how SSM was used to manage the analysis, design, and implementation of purposeful actions to ameliorate AESH and sustainability concerns in the intensive study sites (ISSs).

5.2 PROCESS AND METHODS

5.2.1 EXAMINATION OF THE PROBLEM SITUATIONS

Problems and concerns in the agroecosystem were identified and described during participatory workshops in each of the six study sites. The participatory process is described in Chapter 3; the process of selecting the study sites is described in Chapter 2. Problem, concerns, and issues were termed *messy situations* if there were disagreements on the nature of the problem, its causes, historical background, or potential solutions.

Semistructured interviews were held with groups and individuals in the community having a different perspective on the issue. Table 5.1 shows a checklist of the topics covered in the interviews. Any institution, group, or individual mentioned by interviewees (in relation to the problem situation) were also included in the list of those to be interviewed. The perspectives of each group were captured in rich pictures with different color lines showing agreement or disagreement among various groups or individuals.

5.2.2 ROOT DEFINITIONS AND SOFT SYSTEM MODELING

For each problem situation, primary task and issue-based human activity systems were identified based on the coping strategies and the opportunities mentioned by the individuals or groups interviewed. Root definitions of these systems were then derived and models built to satisfy the basic properties of a system as described by Checkland and Scholes (1990). A root definition was derived for each stakeholder with a different perspective on the problem situation. Different metaphors, based on the roles, norms, and values ascribed to various stakeholders, were used to represent each different perspective on the issue. For each model representing a purposeful

TABLE 5.1
List of Topics Covered in Semistructured Interviews on a Problem Situation

Sequence	Topic	Subtopic
1. Beginning	Historical background	How the problem arose
		Who is/are the most knowledgeable person(s) on the issue
		What are some of the consequences that have been observed
	Nature of problem	Causes
		Effects
2. Mid	Opportunities	How can the problem be addressed
		What resources are needed
		What are the coping strategies
3. End	Stakeholders	Who are/should be the actors in solving the problem
		Who are/should be the beneficiaries
		Who are/should be the owners of the process
		What is the relationship between owners, actors, and beneficiaries
		Who is to blame or are involved in causing the issue
		What has been the role of (named) stakeholder in the situation

action, the monitoring and control unit was identified. Measures of performance, based on what the effects and the causes of the problem were perceived to be, were listed together with their targets and thresholds.

5.2.3 Building Consensus, Compromise, and Collective Action

The rich pictures and models were presented to the different groups or individuals first separately and then together. Participants were asked to comment on the accuracy of the opinions depicted and what the implications appeared to be. Participants were informed that there was opportunity to change any aspect of the models or depictions that represented their own ideas. Changes in the models or depictions were effected, with the participants required to state whether the changes they requested were a change in their opinion (or view or perspective), a compromise, or simply correction of an error in the depiction of their views. Where all participants were present, models were presented as the views and opinions of the facilitator on the problem situation. Criticism of the models and depictions by the community and other stakeholders were therefore directed to the facilitator and not to the group whose ideas were depicted. The identity of the groups or individuals whose views were depicted in a model was not revealed to other stakeholders bearing a different view.

5.3 RESULTS

5.3.1 Problem Situations

5.3.1.1 Drainage and Access Road Problems in Kiawamagira

During the rainy season and every time it rained for a few hours, most of the land in Kiawamagira became flooded. Furthermore, the runoff was too much and had created big gullies in various places in the village and along the only access road to the village. The consequences were that the village was inaccessible at such times, crops were destroyed, and soil productivity was declining. Plate 5.1 is a photograph taken in October 1997 (shortly after the village workshop) showing one of the gullies caused by runoff in Kiawamagira village. Plate 5.2 shows the state of the access road during this time. Plate 5.3 shows one of the outlets passing under the railway line that directs runoff to Kiawamagira village.

Figure 5.1 is a rich picture depicting the various perspectives of different groups within the community and of other stakeholders. There were three main competing perspectives on the causes of flooding, gully formation, and destruction of the access road. The first was that the redirected runoff from the railway and road was the main cause. The course of action, according to this perspective, was to take the institutions involved in the redirecting of runoff to court with a view to compelling them to act. Figure 5.2 shows the root definition and an activity system based on this perspective. The second perspective was that it was the farmers who had redirected the runoff from the farms to the access road, resulting in damage and gully formation along the waterways. Based on this, the course of action was to cooperate as a village and find ways and means of redirecting the runoff away from the village. The root definition

PLATE 5.1 One of the gullies in the Kiawamagira intensive study site, Kiambu district, Kenya, October 1997.

PLATE 5.2 Damaged access road to the Kiawamagira intensive study site, Kiambu district, Kenya, October 1997.

based on this perspective is shown in Figure 5.3. The third was that the area was a swamp before settlement and therefore prone to flooding. The proposed action was therefore to find means of preparing the community to better cope with flooding and damage (Figure 5.4).

5.3.1.2 Gitangu Water Project

According to the participants in the initial village workshop, the Gitangu water project was initiated as a self-help project by the community in Ngecha Sublocation, Limuru Division, in Kiambu district in 1962. The project drew water from Gitangu springs located within Gitangu village and pumped it, using an electric pump, to two water reservoirs in the sublocation, one of which was in Gitangu village. The project was taken over by the Ministry of Water Development in 1965. The ministry was the main manager until 1980, when the pump and other accessories were

PLATE 5.3 One of the outlets (white arrow) passing under the railway line (on ridge) and directing runoff into the Kiawamagira intensive study site, Kiambu district, Kenya, October 1997.

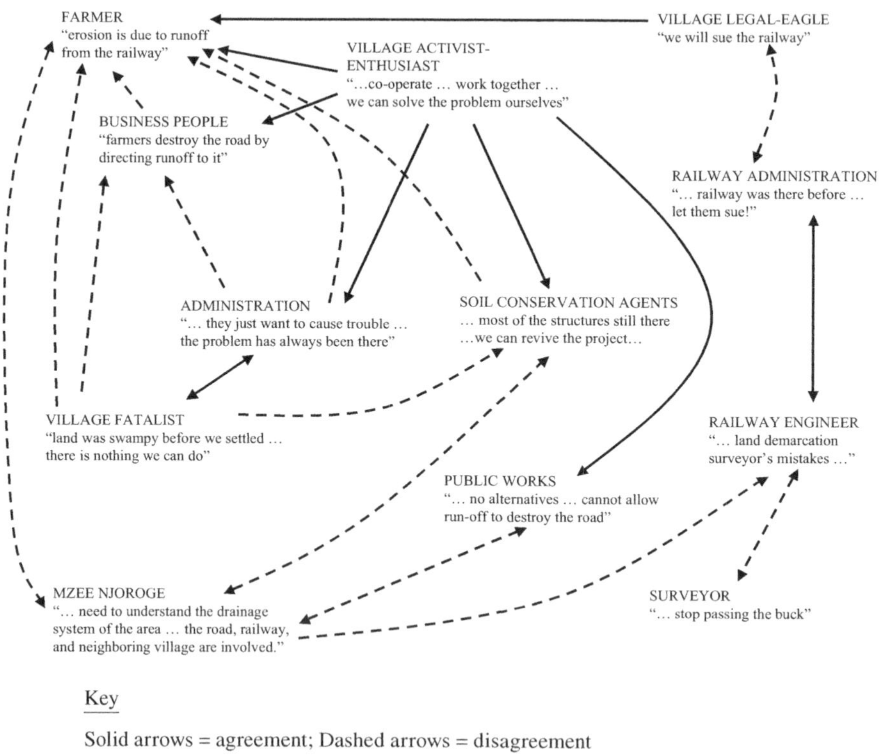

FIGURE 5.1 Differing perspectives on the drainage and access road problems in Kiawamagira village. See CD for color image.

vandalized, leading to the collapse of the project. The project was then handed back to the community, and a committee, including members from the entire sublocation, was selected to revive it. To the time of this study, this had not been successful.

> A system, owned by Kiawamagira community, together with the MoPW and KR to rehabilitate the drainage system in and around Kiawamagira village in order to reduce flooding, gully formation, and access road damage, using resources from the MoPW, KR, and Kiawamagira communities.

C: Kiawamagira community

A: MoPW, KR

T: Runoff directed into village → Runoff directed away from village

W: It is the runoff from the roads and railway that overloads the drainage system of the village

O: Kiawamagira community, MoPW, KR

E: Goodwill from MoPW and KR, Resources from MoPW and KR, community organization, and unity

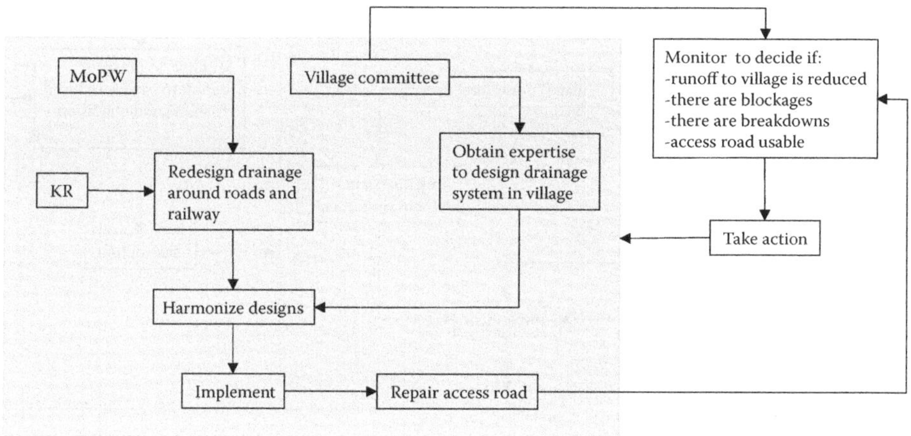

C: Clients; A: Actors; T: Transformations; W: World-view; O: Owners; E: Environment

FIGURE 5.2 A root definition and an activity system based on the perspective of farmers to ameliorate flooding and damage to the access road in the Kiawamagira intensive study site. KR, Kenya Railways; MoPW, Ministry of Public Works. See CD for color image.

Figure 5.5 depicts the differing perspectives of various stakeholders on the Gitangu water project. The three main perspectives with regard to the course of action were (1) do nothing, (2) rehabilitate the project, and (3) start a new project (to sink a borehole).

A system, organized by the Ministry of Agriculture (MoA), together with farmers in Kiawamagira village to design, implement, and maintain proper drainage of farms using resources provided by the ministry and residents of Kiawamagira village.

C: Kiawamagira community

A: MoA, Kiawamagira community

T: Runoff directed to access road → Runoff directed away from access road

W: Farmers direct runoff from their farms to the access road causing damage and flooding in other areas of the village

O: Kiawamagira community

E: Expertise from MoA, Resources from Kiawamagira community, community organization and unity, goodwill from farmers

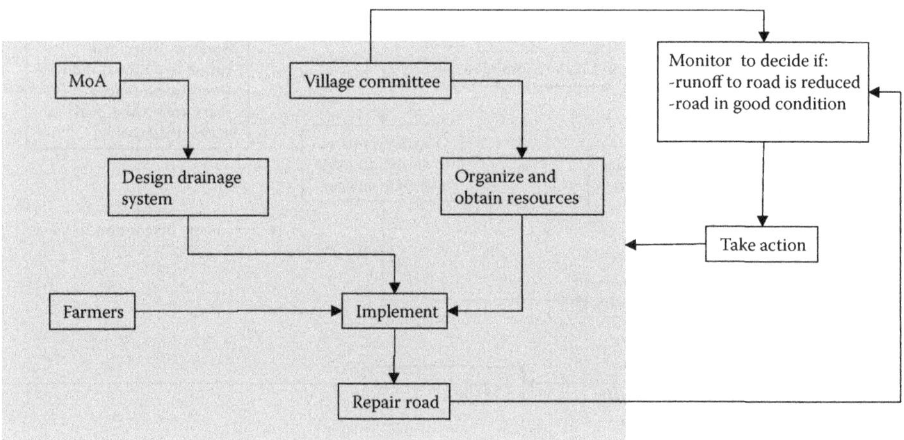

C: Clients; A: Actors; T: Transformations; W: World-view; O: Owners; E: Environment

FIGURE 5.3 A root definition and an activity system based on the perspective of a group of business people and those with off-farm employment in the Kiawamagira intensive study site. See CD for color image.

5.3.1.3 Inadequate Extension Services in Kiambu District

According to participants in the initial workshops, government extension agents visited most of the villages in Kiambu district and demonstrated modern farming techniques. This became less and less frequent beginning from the late 1980s and

A system, organized by the residents of Kiawamagira village, to reduce the negative impacts of flooding and damage to access road during the rainy season using resources available to each of the households in the village.

C: Kiawamagira community

A: Kiawamagira community

T: Suffering resulting from flooding and access road damage → Suffering reduced

W: Lack of adequate preparation during the rainy season increases suffering when the floods come and the access road is damaged
The village is predisposed to flooding and nothing can be done about that

O: Kiawamagira community

E: Resources available to households, community organization

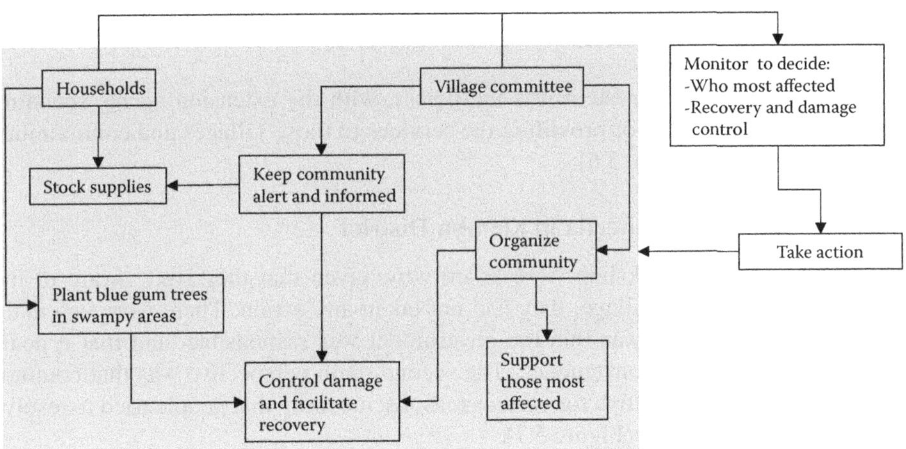

C: Clients; A: Actors; T: Transformations; W: World-view; O: Owners; E: Environment

FIGURE 5.4 A root definition and an activity based on the perspective of a group of community members collectively referred to as the fatalists. See CD for color image.

was almost nonexistent in the late 1990s. There were three main perspectives on the causes of this. The first was that the Ministry of Agriculture could no longer afford to finance such activities. The second was that farmers in many of the villages were too resistant to the extension agents, resulting in disillusionment. The third perspective

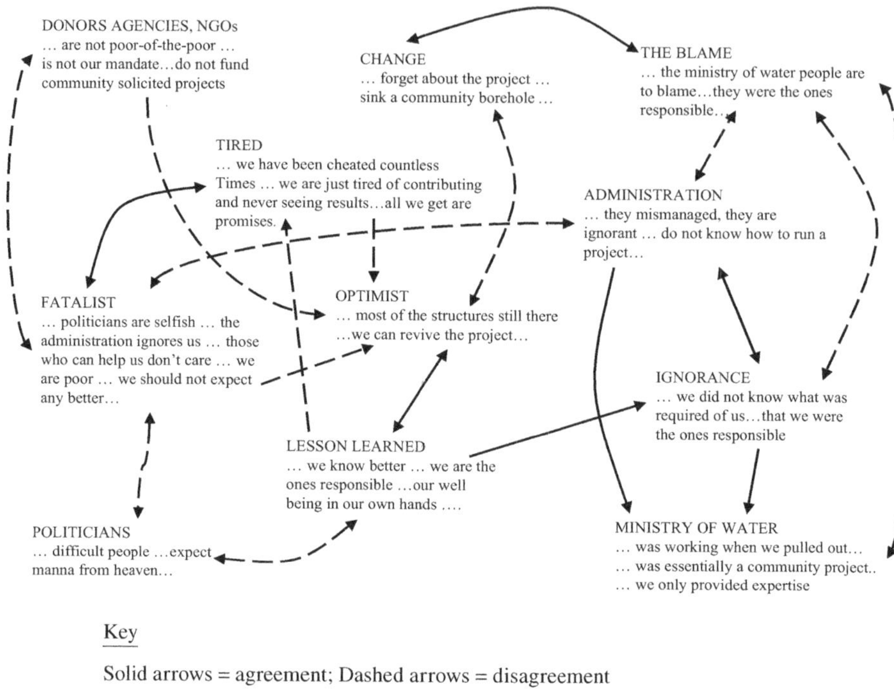

FIGURE 5.5 Differing perspectives on Gitangu water project in Gitangu village. NGO, nongovernmental organization. See CD for color image.

was that the extension system was inefficient, with the extension agents spending time doing other things or providing the services to those villages and communities that they favored (Figure 5.6).

5.3.1.4 Community Inertia in Kiambu District

Participants in the workshop were asked why, given that they were aware of the problems facing their village, they had not taken any action. There were two main perspectives. The first was that the government was responsible, and that it deals with issues at its own convenience. The second main perspective was that community leaders were ineffective for various reasons, meaning that people tried to resolve problems as individuals (Figure 5.7).

5.3.2 Building Consensus and Root Definitions

Figure 5.8 shows the change in perspectives of the various stakeholders on the flooding and access road problem in Kiawamagira village after viewing the completed rich picture of the problem situation. Similar changes in perspectives occurred in all other problem situations in which the approach was used. Table 5.2 shows the activities, measures of performance, and targets negotiated to resolve four of the problem situations faced by communities in the six ISSs. In the situations, communities began the implementation process immediately after the first rich pictures were

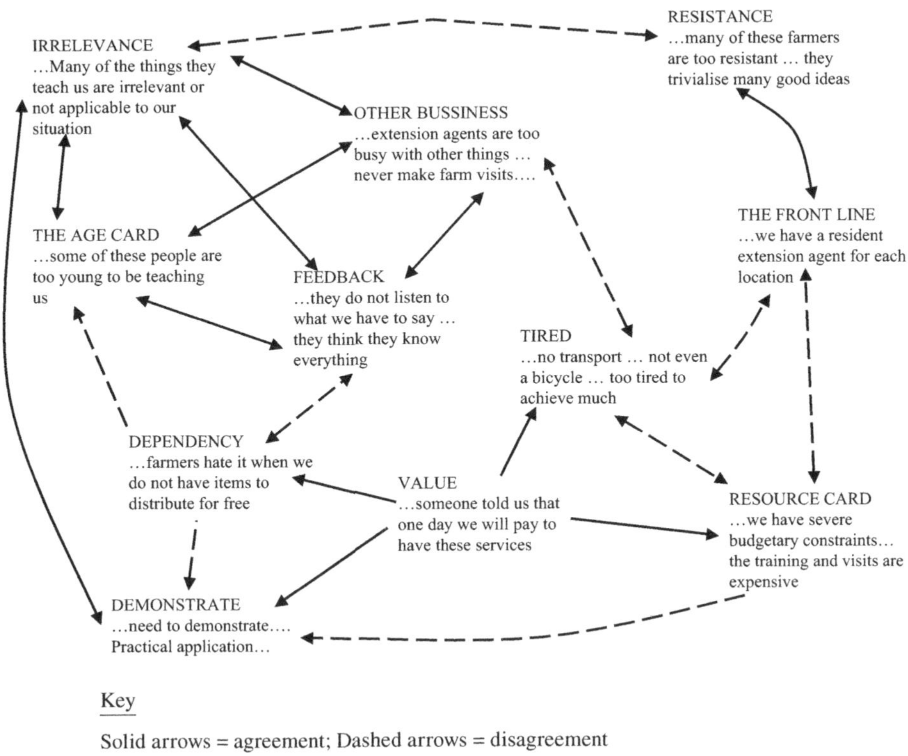

FIGURE 5.6 The perspectives of various stakeholders on the causes of inadequate extension services in the district. See CD for color image.

drawn, proceeding based on their action plans. In the case of Kiawamagira village, the activities undertaken to repair the access road were followed by severe drainage and gully formation, resulting in an even worse road condition. Plate 5.4 shows the condition of the road in November 1997, at 1 month after the initial workshops. Stones from the nearby quarry were put in place to fill the gullies, but the runoff was still directed toward the village and its access road.

5.3.3 Implementation, Monitoring, and Evaluation

In Kiawamagira village, negotiations were carried out with the Kenya Railway and with the Ministry of Public Works in Kiambu. Both pledged material support as well as expertise as requested. The community was to raise the initial funds to start the work. The AESH project provided an engineer, who designed a drainage system. The Ministry of Agriculture provided staff and support in the management of soil erosion and gullies. The implementation of the drainage system has not yet been implemented. This was attributed to the inability to raise the required funds, the death of two of the key committee members, and the emigration of the committee chairman. Attempts to raise funds and implement the project are still ongoing. Plate 5.5 shows the success of the soil conservation measures on one of the gullies in the village.

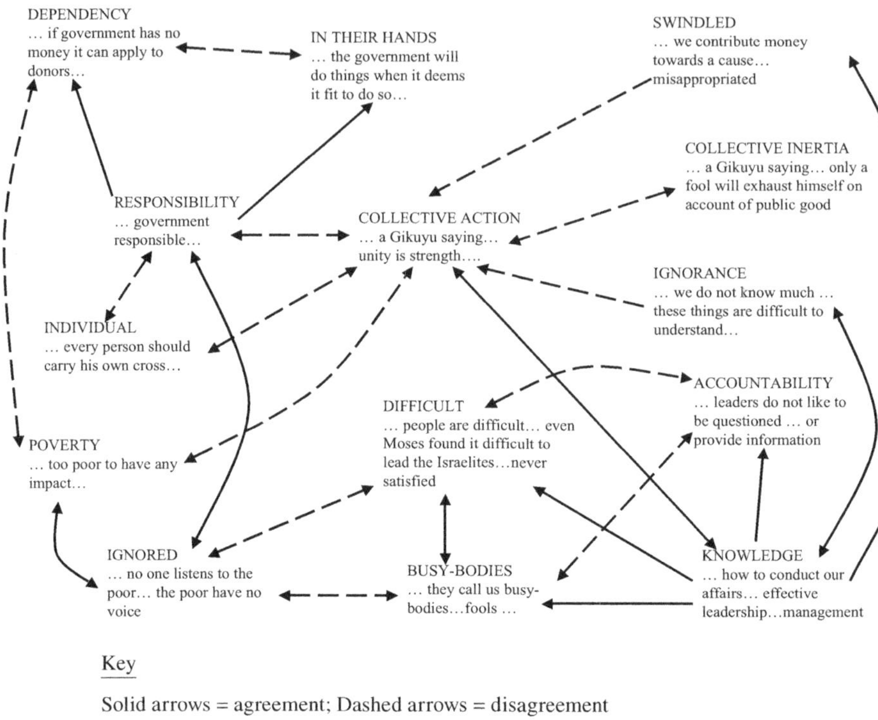

FIGURE 5.7 Perspectives of community leaders in the intensive study site on the inability to act on problems affecting their agroecosystems. See CD for color image.

In Gitangu village, the community raised 800,000 Kenya shillings and purchased a new pump. They obtained further financial support from Plan International to purchase a booster pump, build a bigger tank, and rehabilitate the piping system. Water kiosks are being built at strategic points in the village, while plans are under way to purchase water meters and to provide piped water to most homesteads. Plate 5.6 shows community members preparing the site for the new water tank. In the background is the old water tank that was to be rehabilitated and used as the treatment unit for the new water supply system. Plate 5.7 shows work being carried out to rehabilitate the piping system.

Communities in all six villages organized meetings with extension agents to discuss various topics. In most of the cases, the meetings were organized and funded by the communities, but in some cases, the initiative came from the extension agents following the presentation of community action plans to the divisional extension staff. In all cases, topics to be covered were selected in consultation with the communities. Plate 5.8 shows an extension agent demonstrating the use of various energy-saving devices in Kiawamagira village.

Plate 5.9 shows a group of leaders from the six ISSs at the end of a 6-day residential training workshop on leadership and community mobilization together with some members of the AESH multidisciplinary team. At this workshop, leaders developed

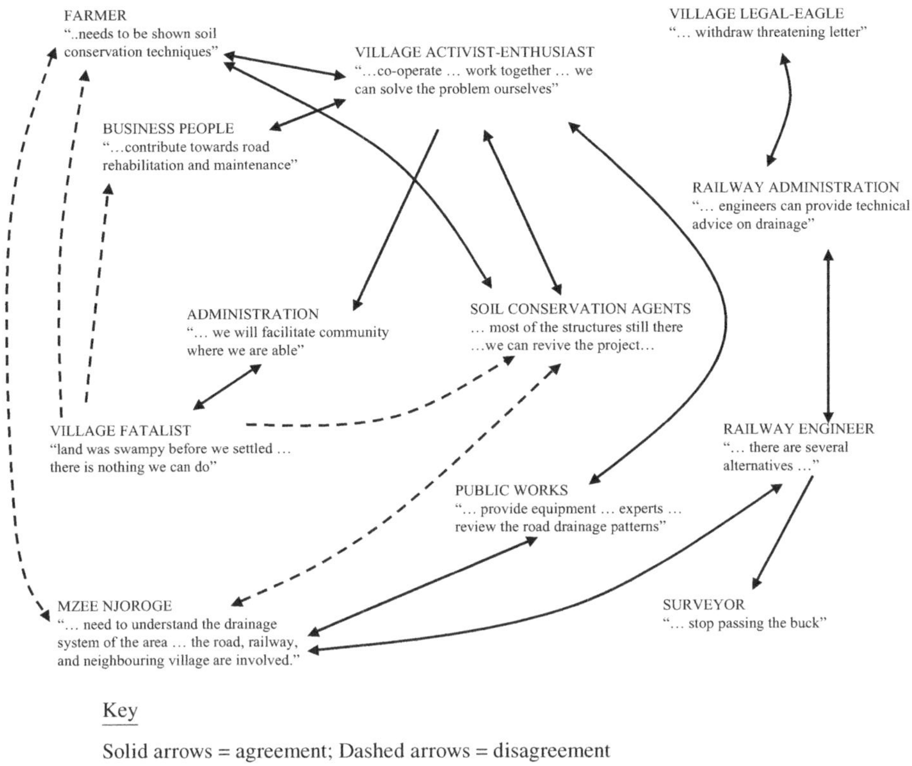

FIGURE 5.8 Changing perspectives of various stakeholders on the flooding and access road problem in Kiawamagira village. See CD for color image.

the intervillage monitoring and evaluation program. Plate 5.10 shows an intervillage evaluation meeting in Githima village. Included in this meeting were officials of the International Institute for Rural Reconstruction (IIRR) as observers.

5.4 DISCUSSION

5.4.1 Soft System Models

The rich pictures were instrumental in helping communities to see the different perspectives that existed on a problem situation. In many cases, they elicited laughter and a softening of stance by the various protagonists. By creating root definitions based on the various perspectives, researchers were able to make communities focus on the strength and weakness of each perspective, thereby generating an opportunity for synthesis, negotiation, and compromise.

The main difficulty with the process is that it took time to build the rich pictures and root definitions, while the communities preferred to start with the implementation process almost immediately after the initial village workshops. Many of these attempts resulted in failure and frustration on their part. How the process can be

TABLE 5.2
Activities, Measures of Performance, and Targets Negotiated and Agreed On to Resolve Four Problem Situations Faced by Communities in the Six Extensive Villages of Kiambu District, Kenya

Problem Situation	Flooding and Access Road Damage in Kiawamagira	Gitangu Water Project	Inadequate Agricultural Extension Service	Community Inertia in Kiambu District
Transformation	Reduce soil erosion, flooding, and access road damage	Provide clean, potable water to households in Gitangu village	Make extension services accessible and gainful to the communities	Empower communities to solve their problems
Process	Create a drainage system	Rehabilitate Gitangu Water Project	Create a cost-sharing and feedback system	Increase the communities' capacity for collection action
Owners and actors	Kiawamagira community Railway Public Works Ministry of Agriculture	Gitangu community	Communities Government extension staff	Communities Agroecosystem health project
Activities	Elect a committee Study drainage pattern Design methods to reduce flooding Look for funds Build drainage system Put gravel on road Maintain the drainage system	Elect a committee Obtain water permit and title for the Gitangu springs Look for funds Purchase pump and reconnect electricity supply Rehabilitate the piping system Build water Kiosks	Make schedules of topics to be covered Discuss with extension staff on the calendar schedule and materials required Decide on a cost-sharing scheme Have workshops Provide feedback at end of workshop	Elect committees Have leadership training workshops Hold regular participatory meetings Continuous monitoring and evaluation of projects Feedback by leaders to communities Feedback by community to leaders

Measures of performance and their ranges	Gully formation (none, many new ones) Gully progression (reversed, increasing depth and head) Frequency of flooding (1/5 years, twice/year) Status of road (very good, impassable)	Average distance to water source (< 200 m, > 1 km) Average expenditure on water/day/household (< 20 sh, > 200 sh)	Number of extension workshops held/year (none, > 12)	Number of meetings/year (none, > 12) Attendance (% of households) (10, > 90)
Targets	No new gullies Reverse progression of new ones Flooding less than once every 5 years Road passable throughout the year	Water source less than 200 m for each household in village Average expenditure on water per day/household < 20 sh	Extension workshop once every month	Participatory meeting once every month Minimum attendance by 80% of households

PLATE 5.4 The condition of the access road in Kiawamagira in November 1997 after the initial attempts by the community to repair it.

PLATE 5.5 A healing gully in the Kiawamagira intensive study site after community intervention with assistance from the Ministry of Agriculture, Kiambu district, Kenya, January 1999.

incorporated into the initial process is worth considering in order to guide communities toward activities that are more likely to succeed. In the context of action research, initial failures can be viewed as learning experiences that result in creation of local theory on project implementation. However, where resources are limited and the communities' capacity for collective action is weak, it is likely that initial failure may result in further degeneration into community inertia. In Kiawamagira village, initial rehabilitation was followed by massive gullies along the access road, but the communities were not deterred from trying. They recognized that they had not assessed the situation adequately and hence causing the failure. In Thiririka, attempts were made to develop a water project from a natural spring in the village. It was found that this spring could only have an output of not more than 80 liters of water a day. This disheartened the community to the point that only nominal attempts have been made to implement most other activities in their action plans.

PLATE 5.6 The old water tank and the preparation of the site for the new water tank in the Gitangu intensive study site, Kiambu district, Kenya, June 2000.

PLATE 5.7 Work being carried out to rehabilitate the piping system of the Gitangu water project, Kiambu district, Kenya, June 2000.

5.4.2 Collective Action, Action Research, and Soft System Methodology

While the soft system approach was instrumental in generating syntheses or negotiated goals and objectives, there is need for further evaluation of whether these lead to more sustainable project implementation and enhanced collective action. Initial indication is that this is the case. It may also be that public debate about various viewpoints generates a need to present a consensus while the actual positions are largely unchanged. These would arise later in the form of leadership wrangles or lack of participation by some groups in the community. When linked with action research, the approach provides opportunities for review and remediation.

By listing the activities, the expected transformation, the measures of performance, and the targets, this approach provides a means for evaluating progress.

PLATE 5.8 An extension agent from Kikuyu Division demonstrating the use of various energy-saving devices during a workshop in the Kiawamagira intensive study site, Kiambu district, Kenya, June 1998.

PLATE 5.9 Community leaders from intensive study sites and some research team members at the end of a 6-day residential training workshop on community leadership and mobilization together, Waruhiu Farmers Training Institute, Kiambu district, Kenya, May 1998.

Iterative steps of implementation, monitoring, and reflection allow for short-term planning toward medium-term and long-term goals. Activities can be reviewed in the face of changing circumstances such as new opportunities, new knowledge, or lack of resources.

PLATE 5.10 Members of the agroecosystem health village committees from all the intensive study sites reviewing the outcome of an intervillage evaluation meeting together with officials of the International Institute of Rural Reconstruction (IIRR), Githima intensive study site, Kiambu district, Kenya, May 2000.

REFERENCES

Bulow, I.V. (1989). The bounding of a problem situation and the concept of a system's boundary in soft systems methodology. *Journal of Applied Systems Analysis* 16: 35–41.

Checkland, P., and Scholes, J. (1990). *Soft Systems Methodology in Action.* Wiley, Chichester, U.K. 329 pp.

Checkland, P.B., and Wilson, B. (1980). Primary task and issue-based root definitions in systems studies. *Journal of Applied Systems Analysis* 7: 51–54.

Flood, R.L., and Carson, E.R. (1993). *Dealing with Complexity: An Introduction to the Theory and Application of Systems Science.* New York: Plenum Press.

Ison, R.L., Maiteny, P.T., and Carr, S. (1997). Systems methodologies for sustainable natural resources research and development. *Agricultural Systems* 55: 257–272.

6 Development of Health and Sustainability Indicators for a Tropical Highlands Agroecosystem

6.1 INTRODUCTION

Describing agroecosystems, assessing their sustainability and health, and assessing progress toward community goals and objectives has become of great interest to researchers, development agents, and communities. The agroecosystem health approach proposes that these descriptions and assessments can be achieved using a group of carefully chosen indicators (Rapport and Regier, 1980; Gosselin et al., 1991; Lightfoot and Noble, 1993; Rapport, 1992; National Research Council, 1993; Cairns et al., 1993; Izac and Swift, 1994; Winograd, 1994; Dumanski, 1994; Rapport et al., 1985; Ayres, 1996; Smit et al., 1998). There are numerous definitions of what constitutes an indicator (Boyle, 1998; Boyle et al., 2000). Gallopin (1994a) and Smit et al. (1998) described *indicators* as measurements that can be taken for a given complex phenomenon to document how it changes over time, how it varies across space, and how it responds to external factors. In terms of an agroecosystem, an indicator has been defined as a measurable feature that singly, or in combination with others, provides managerially or scientifically useful evidence of ecosystem status (Canadian Council of Ministers of the Environment [CCME], 1996) relative to a predefined set of goals.

Selection of indicators is complicated by two main difficulties. First, the list of potential indicators varies from one agroecosystem to another as well as among levels in an agroecological hierarchy. The second difficulty is that there are virtually an infinite number of measurable parameters at each hierarchical level of an agroecosystem (Schaeffer et al., 1988). There are, however, some important guidelines in the selection of agroecosystem indicators. A systems approach should be taken to select a comprehensive set of measures. In addition, the choice of indicators must be explicitly guided by societal issues and values (Kay, 1993) that give meaning to the description or assessment process. This ensures that selected indicators are practically useful in terms of decision making, setting policy guidelines, or scientific research. It can be argued that some "nonquantifiable" indicators provide more important information than more objective ones (Harrington, 1992). But, if the objectives are to assess the direction or magnitude of change in the status of agroecosystems, to compare one system with another, or to assess the potential impact of various strategies and management options, then indicators must be amenable to an objective assessment. Selection of indicators must also be tempered by practicality and the cost of measurement in terms of time and money.

The CCME (1996) proposed a framework through which a suite of health and sustainability indicators can be developed. First, a systemic description of the ecosystem under review is developed using a variety of methods, including participatory approaches. Essential components of a systemic description of an agroecosystem are goals and objectives of the human communities living in them and a definition of what constitutes health for that agroecosystem. Indicators are then selected based on identified health attributes, community goals, objectives, and values, and are guided by a list of desired qualities for an indicator.

Under this scheme, categories of measures that reflect the goals and values of the system are generated. Within each category, measures for which data can be practically obtained are identified as potential indicators. The choice of a measure in an initial list of indicators depends on its desired qualities as an indicator. Such qualities include *validity*, which is the degree to which an indicator reflects changes in the system (Dumanski, 1994); cost-effectiveness, timeliness; sensitivity; and ease of measurement (CCME, 1996; Smit et al., 1998). Casley and Lury (1982) listed five considerations when selecting indicators. (1) Can it be unambiguously defined in the conditions prevailing? (2) Can it be accurately measured in the conditions prevailing and at an acceptable cost? (3) When measured, does it indicate the state of the agroecosystem in a specific and precise manner? (4) Is it an unbiased measure of the value of interest? (5) When viewed as one of a set of indicators to be measured, does it contribute uniquely to explaining the variation in health and sustainability?

Initially, a large number of variables meeting these criteria may be included in the list of indicators. However, many of the variables first selected are unlikely to provide important additional information relative to other variables in the group. Thus, statistical and mathematical methods to develop useful subsets of indicators can be very helpful in developing suites of indicators that optimize parsimony and information provided. Such methods include principle components analysis and multiple correspondence analysis (MCA). This chapter describes how a group of indicators of agroecosystem health and sustainability was developed for use in a tropical highlands agroecosystem and an evaluation of their practicality and application.

6.2 PROCESS AND METHODS

The objective was to develop a suite of indicators suitable for use by researchers, policymakers, and communities to assess the health and sustainability of the Kiambu agroecosystem. Two broad approaches were used. The first was a participatory process involving communities in the agroecosystem. Indicators developed in this process were referred to as *community-driven indicators*. The second approach derived lists of potential indicators from the stated agroecosystem problems, needs, objectives, and goals and from suggestions—by a multidisciplinary team of experts—of variables that were felt important. These were referred to as *researcher-proposed indicators*. Figure 6.1 is a conceptual framework of the process used in this study to develop suites of agroecosystem health and sustainability indicators.

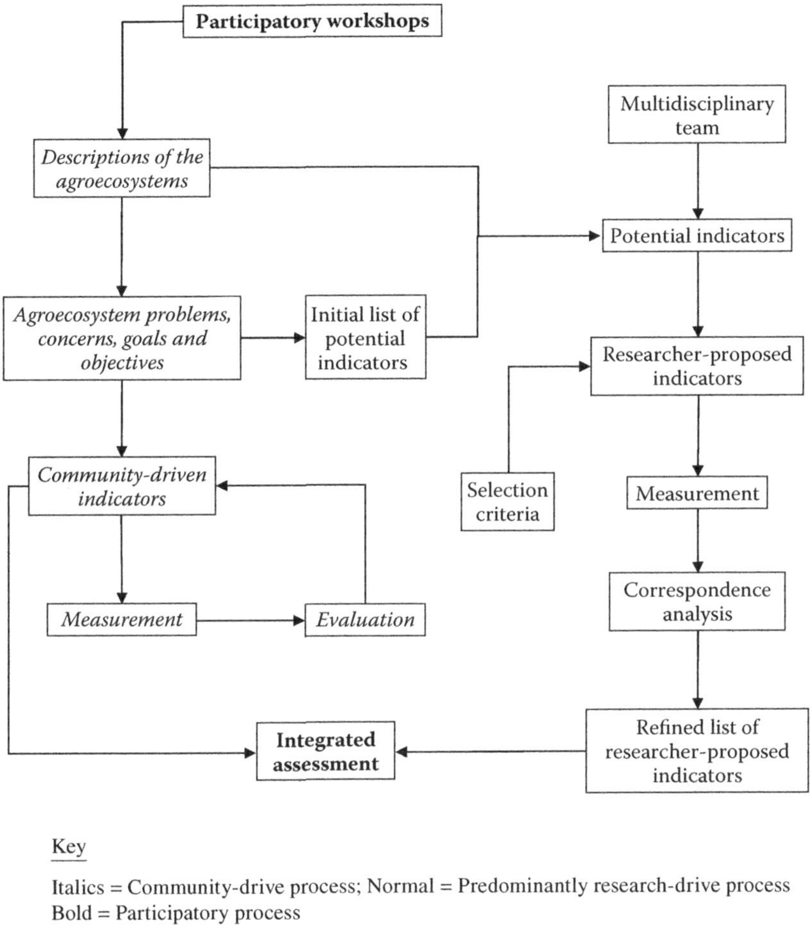

FIGURE 6.1 Flowchart showing the approaches in which indicators of agroecosystem health and sustainability were developed.

6.2.1 Development of Community-Driven Indicators

The rationale for developing community-driven indicators was that communities must assess their own agroecosystems for the process to be sustainable. However, indicators selected by researchers may not be practical for use by the communities. Communities in the six intensive study sites were facilitated to develop a suite of indicators that they would use to monitor the health and sustainability of their agroecosystems. These indicators were developed in 3-day workshops held in each of the six intensive villages in July to August 1998. Gender- and age-specific focus group discussions were used in conjunction with pairwise ranking and trend analysis to identify health attributes of most concern to the residents, list potential indicators, and then refine the list to a parsimonious suite. The sequence of participatory tools

TABLE 6.1
Sequencing of Learning Tools Used to Generate Community-Driven Health and Sustainability Indicators

Tool	Objectives	Output
1. Introduction and icebreakers	Develop rapport Explain workshop objectives Workshop logistics (venue, meals, schedule)	List of participants by gender Workshop logistics
2. Focus groups Topic: "Monitoring and Evaluation"	Introduce concepts (monitoring, evaluation, and indicators)	Definitions of *monitoring* and *evaluation* Understanding of indicators
3. Focus groups Topic: "Ecosystem Health"	Introduce concept (ecosystem health) Describe a hypothetical healthy ecosystem Define agroecosystem health	Understanding of ecosystem health Identification of some health attributes
4. Group presentations	Identify disparities among groups on the definition and conceptualization of ecosystem health	Understanding of ecosystem health
5. Listing ecosystem health attributes	Identify ecosystem health attributes	Lists of attributes
6. Pairwise scoring matrix	Rank attributes based on their role in determining ecosystem health	Rank matrix of attributes
7. Focus groups Topic: "Indicators of Ecosystem Health	Identify potential indicators for selected health attributes	Lists of potential indicators
8. Group presentations and scoring matrices	Assess selected indicators in terms of validity, ease of measurement, and usefulness	Refined lists of health indicators
9. Planning for ecosystem health monitoring	Identify resources and people to carry out ecosystem health monitoring using selected indicators	Itinerary of an ecosystem health-monitoring activity

used in these workshops and their objectives and expected outputs are shown in Table 6.1. Details of the specific tools used are provided in Chapter 2.

After explaining the objectives of the workshop and seeking the communities' consent, the concepts of indicators, monitoring, and evaluation were introduced through focus group discussions. To introduce the concept of indicators, participants were asked to reflect on their stated agroecosystem goals as well as their concerns or problems and to find things that they would measure to find out if there was an improvement. Health was equated to the Gĩkũyũ term *ũgima* which is used interchangeably to mean unity, maturity, and wholeness. It is also used with reference to a human being to mean either a mature, well-rounded person or a healthy (broadly defined) person.

Participants were asked to describe their vision of a healthy village. They were then asked to list the likely negative consequences of current activities, processes or states in the village that threatened this vision. Discussion on what could be done to increase the chances of realizing the vision of a healthy village followed, with the facilitators introducing an individual's health as an analogy. Once the participants agreed on the value of self-assessment, focus group discussions were initiated to discuss (1) what indicators (*ithimi*) are, (2) why indicators are useful, (3) which ones would be most relevant for the particular village, (4) how empirical measurements (*gũthima*) would be carried out, and (5) how this information would be used.

Each group presented their conclusions to a joint forum, and further discussion was encouraged. Disparities and points of agreement among groups were noted. Participants were then asked to list those attributes that they felt were the most essential elements of agroecosystem health. Pairwise scoring was used to rank attributes in terms of importance. Focus groups were then reconstituted and each asked to list potential indicators for the 10 most important health attributes identified. Communities were encouraged to consider both the practicality of measuring a given indicator and its validity.

6.2.2 Development of Researcher-Proposed Indicators

The researcher-proposed indicators were based on the descriptions provided by the communities through the participatory process, their stated goals and objectives, and the attributes they considered to be most influential to agroecosystem health and sustainability and depicted in cognitive maps. The initial list of potential research-proposed indicators was arrived at using two different methods. In the first method, lists of potential indicators were generated from the cognitive maps and community goals. A potential indicator was a measure that would reflect an important change in the potential of the system to meet a stated goal or one that reflects an important change in a problem situation. An initial list of potential indicators was generated combining all the goals and concerns from the six study sites.

The second method of generating potential indicators was through suggestions by experts from various disciplines. In this process, the descriptions provided by the communities through the participatory process as well as the initial list of potential indicators derived from agroecosystem problems and goals was provided to a team of experts consisting of social scientists, veterinarians, agriculturalists, engineers, and medical professionals among others. The experts then proposed indicators that, they felt, would provide important information in addition to that provided by variables in the initial list.

Indicators were selected from the list of potential indicators based on (1) validity, (2) feasibility, (3) parsimony, (4) timescales in which changes were reflected, (5) holarchical scales at which measurements can be taken, and (6) ease of interpretation. *Validity* was defined as how well a variable reflected changes of the attribute it was intended to measure. *Feasibility* was defined as the practicality of measurement (technical feasibility) and the cost (in terms of time and other resources) of measuring a given variable (economic feasibility). The principle of parsimony was included as a criterion because some variables provided information on more than one attribute. For parsimony, some variables were excluded for the suite without any significant loss in amount and quality of information supplied by the indicators. Those variables that were not feasible to measure at the targeted holarchical scales were not included. In addition, indicators were categorized based on the scale at which they could be measured or interpreted.

In the initial suite of indicators, validity, feasibility, and parsimony were assessed qualitatively. The time- and holarchical scales were based on the target timescales and holarchical levels on the entire health and sustainability assessment. Ease of interpretation was assessed by listing all the likely outcomes for a particular variable (if discrete) or a range (if continuous) and stating what the conclusions would be for each likely outcome or extreme in a range. If the conclusions were equivocal, then an indicator was considered unsatisfactory in terms of interpretation.

6.2.3 Indicator Measurements

6.2.3.1 Community-Driven Indicators

Measurement of community-driven indicators was community based and in the form of participatory monitoring and evaluation. This was based on the assumption that such an assessment provided stakeholders with information crucial for the successful management of the agroecosystem. In each of the six intensive villages, indicators were divided into 8–10 sets (each with four to six indicators). Groups of 8–10 community members were then formed, and each was assigned a set of indicators to measure (gũthima). The village agroecosystem health committee was assigned the coordinating role. Regular (twice-a-week) group meetings were scheduled for a period of 1 month for this purpose. A village participatory workshop was held at the end of this period; analyses of the information gathered were conducted at these workshops.

6.2.3.2 Researcher-Proposed Indicators

An initial empirical assessment was made using the initial suite of indicators. Indicators were categorized based on the methods (questionnaire, laboratory tests of samples, participatory methods) to be used for its measurement and the scale at which it would be measured (village or land-use units). For indicators to be measured using a questionnaire, a relational database was created using Microsoft Access. Indicators to be measured using a questionnaire were entered in a table that was linked to a set of tables that contained the questions, their choices (if structured), and the data categorized by level. The questionnaire was generated from the tables using filters and sorting procedures to prevent duplication of questions and information and to provide a logical flow. Three teams of two people each (from the research team) were trained on the questionnaire and its objectives to enable them to administer the questionnaire. The questionnaire was pretested on a random sample of farms (four in each village) and changes made based on the recommendations of the teams and the interviewees.

For measurement at the land-use level, 20 land-use units were selected from each of the six study sites. The units were selected at random from a list of all the land-use units in the village. Owners were contacted for permission to participate in the study. Dates and times for the interviews were set based on the availability of the interviewees. The allocation of interviewees to each of the three teams of interviewers was randomized. For land-use-level indicators that required laboratory testing, samples (water and soil) were obtained from the same units in which the questionnaire was applied. Participatory methods used to measure some of the indicators at village level are similar to those described in Chapter 3.

6.2.4 Refining Researcher-Proposed Indicators

Multiple correspondence analysis was carried out using the PROC CORRESP of SAS statistical software (SAS Institute Inc., SAS Campus Drive, Cary, NC 27513). A dimension with a significant χ^2 value was interpreted as an attribute of farms/homesteads that, if measured, would explain a significant amount of variation among them. Clusters of factor levels on either extreme of a dimension were examined to enable researchers to ascribe a physical-world term to the attribute represented by a dimension ("reification"). Only variables with factor levels that contributed a significant amount of variation were included in the refined list of indicators. The refined set of indicators was used, in conjunction with the community-driven set, in subsequent assessments of the agroecosystem.

6.3 RESULTS

6.3.1 Community-Driven Indicators

The concepts of health and indicators as applied to agroecosystem were understood and adopted by the communities. Communities accepted the notion of using indicators to assess their agroecosystem. Descriptions given during the indicators workshops indicated a common vision of a healthy community across the six villages. A retired teacher, whose only source of livelihood now is a small-scale farm in Gitangu village, aptly captured this vision:

> We would be having sufficient management skills to run our farms efficiently. We would use simple technologies to reduce the drudgery in farming and daily life. Although farm sizes may still be small, we would have technologies for *scientific fertility farming* [his translation] such that yields would be much higher than the current. Yet the negative impacts on the soil common in our farms today would be minimal.
>
> *People's dependence on government's support would be minimal. We would have enough know-how and resources to obtain services either as a group or privately. We would have enough management skills to run our own community projects effectively.*
>
> Poverty is the greatest enemy in one's life *[his translation] and the only way to deal with it is through knowledge and hard work. ... But an individual's prosperity is meaningful only if the people around him are also prospering. While one person seeks to provide me with enough, clean water, I in turn would seek to provide others with a wholesome food-crop and at a fair price. The other person provides us with transport and so forth so that each ones' needs are met in the best way possible.*
>
> *... Our children would excel in all they do because they would be well fed and healthy. They would realize their full potential in all they do because they would have a secure livelihood to retire to in their old age.*

Communities gave varied answers to the question: How would one tell if this village is getting healthier? Reduction in poverty, increasing wealth, and increasing human health were some of the criteria given by some of the participants in some villages. In five of the six villages, no consensus was obtained on this issue. The workshop in Gitangu village, the first indicators workshop to be held, was the only one to reach an autonomous consensus. The debate was as follows:

Participant 1: In my group, we agreed on how we could tell if our village is becoming healthier. We agreed that if we have plans as a community, and those plans are being implemented properly, then our village is headed towards a more healthy status.

Participant 2: But even thieves and conspirators have plans and they succeed. … sometimes more often than not.

Participant 1: But their actions are harmful. Everybody can see that!

Participant 3: It is not easy to detect negative effects of some of our actions. When you are cultivating, it is a good thing because you get a harvest. But quite imperceptibly your soil keeps deteriorating. Some of it is slowly carried away by runoff. You will not know until many years later. In any case, people are likely to complain even when a good thing is happening. A good example is when a doctor prescribes an injection for your child. You help in restraining the child, and you know it is a good thing. But that does not stop the child from complaining. Does it?

Participants: Of course not! The child will cry.

Participant 2: I think being aware of the consequences of our plans and actions and being ready to deal with them is a very important component of the health process.

Participants: That is very true.

This description was offered to participants in all the workshops and a supplemental question was asked: How can we determine the consequences of plans and actions? Participants used the terms *Kuona mbere*, *Gũikia maitho kabere*, and *Gũthima* to describe the processes. The first two terms translate roughly to projection into the future or prediction, (direct translation: "seeing into the future" and "throwing eyes ahead," respectively). The third term translates to "measuring" or "monitoring" and is also used to refer to the procedures that are carried out before a doctor makes a diagnosis. The following excerpts from the village workshops illustrate the context in which these terms were used and the communities' understanding of indicators.

> We need to know—and prepare for—the consequences of our actions by projecting into the future [Gũikia maitho Kabere]. For example, if we were to continue with our current rate of land subdivision we better start learning how to make storied buildings.

> In the history of this village [Gitangu] [there is] a record of what we are talking about. During the 1956 land demarcation, our forefathers had seen into the future [Kuona mbere]. Of their own consideration, they decided to spare some land for a cemetery in the village. There were no dairy cattle then, and no one in the village had the need for a dip, but they spared some land for a dip. They had no teachers, and only a few of them

> sent children to school. But they spared some land for a school. None of them were buried in the cemetery, and the cattle dip was never built until 15 years ago. Today, there is no one in this village who has not benefited directly or indirectly from their foresight. We wish to do the same for our future and the future of generations to come. We need to assess [Gũthima] the effects of our actions today to make better decisions for the future.

The process of indicator measurement was therefore referred to as gũthima and indicators as ithimi. The value that an indicator takes correctly fitted the term gũthimo. These terms are used in similar contexts in reference to human health and were therefore assumed to be readily understandable by most people in the villages. Participants were then asked to make lists of indicators that they would use to assess specified agroecosystem attributes. These attributes were (1) soil fertility and farm productivity; (2) pests and diseases; (3) environmental quality; (4) incomes, savings, investments, and employment; (5) lifestyle; (6) leadership and community action; (7) knowledge, information, and education; (8) markets and marketing; and (9) equity. Table 6.2 gives a summary of indicators selected for each village.

6.3.2 Researcher-Proposed Indicators

The measured attribute, the categories, and the number of researcher-proposed indicators in each of the three domains are shown in Table 6.3. Most of the categories in the social domain had no indicators mainly due lack of conceptually valid measures of the attributes as well as difficulties in measurement. For the biophysical and economic attributes with no indicators, the main reason was the cost and difficulty of measuring them. Researcher-proposed indicators were divided into two sets based on the level of the agroecosystem holarchy at which they were to be applied. The first set consisted of measures to be applied at the land-use unit (LUU) level, while the other was to be applied at the study-site level (SSL).

A list of researcher-proposed LUU-level indicators is shown in Table 6.4. For profitability and cost scores, indicator crops were coffee, tea, maize, kale, beans, and potatoes. For the preference scores, indicator common foods were maize, beans, peas, kale, carrots, and Irish potatoes. Indicator traditional foods were arrowroots, sweet potatoes, cassava, millet, and sorghum. Indicator resources for equity assessment were land, vehicles, livestock, cash crops, food crops, household goods, children, nonfarm income, and cash savings. Indicator infrastructure included market, public transportation, schools, health care facility, and administrative offices (Appendix 2). Adults were defined as non-school-going persons over 18 years of age. For the purpose of child health clinic (CHC) records, children were defined as those LUU members 5 years of age or younger. Available labor was defined as the total number of adults in the LUU with no off-farm employment. Nonfood crops included traditional cash crops such as coffee, tea, and pyrethrum. Food crops included vegetables, maize, beans, and the like, even when grown primarily for sale. For contacts and familial ties, only visits outside the district were considered.

Table 6.5 is a list of researcher-proposed SSL indicators of health and sustainability for the Kiambu agroecosystem. Most of these indicators were aggregates of measurements taken at the LUU level. Indicator crops, foods, and resources were as

TABLE 6.2
Village-Level Community-Based Agroecosystem Health Indicators, Kiambu District, Kenya, June 1998

Attribute	Mahindi	Kiawamagira	Gitangu
Lifestyle	1. Number of people with proper personal hygiene 2. Types of diets 3. Dress habits	1. Farming techniques: new versus old 2. Types of houses	1. Personal hygiene 2. Types of crops and livestock 3. Time usage
Social organization	Number of completed community projects Number of people attending meetings	Frequency of meetings Number of community plans executed Number of people gainfully employed	Number and severity of needs in the community Number of needs met over the past year
Equity	Distribution of work by age and gender	Meeting attendance by age and gender Distribution of chores, household incomes Unfair cultural practices	Distribution of leadership positions by gender and age
Quality of environment	Distance to water Coloration of water Smell of water	Frequency of waterborne diseases Air quality (bad odors) Personal and homestead hygiene Garbage dumps in public places (road, river)	Types of chemicals used on farm Storage of chemicals in homestead Disposal of containers
Soil fertility	Color of soil Types of weeds	Quantity of harvest Soil color and texture Types of weeds	Soil erosion measures by farms Number of livestock per farm Quantity of harvest taken to market
Farm productivity	Number of homesteads with granaries Expected yields of crops	Types and quantity of foods bought from market	Quantity of produce sold versus purchased
Pests and diseases	Number of hospital visits Number of livestock deaths	Human mortality Human morbidity	Human morbidity

Gikabu-na-buti	Thiririka	Githima
Types of crops planted Variety of items in the market Types of buildings Number of people working outside village	Types of houses Types of crops and livestock Food habits	Food habits Types of crops Types of employment Types of houses
Number of community projects in the village Attendance at meetings Frequency of conflicts in the village Frequency of social contacts between households	Number of community projects completed Frequency of meetings and attendance Frequency of interactions between households	Frequency of meetings in the village Number of projects completed
Proportion of female leaders Distribution of farming labor by gender Distribution of farming resources by age	Proportion of female leaders Youth unemployment	Ownership of resources by gender and age Attendance of meetings by gender and age Distribution of chores by gender and age
Water quality Presence of fish in river	Disposal of agrochemical and related materials Location and use of toilets Location of wells	Frequency of diseases associated with poor environment
Crop yields Number of livestock Number of trees (tree cover)	Remnant of plant materials in the soil Crop yields	Types of weeds growing Gully formation Yellowing of crops
Quantities of produce taken to market Types and quantities of purchases	Milk yield Kale yields per acre	Yield per acre Causes of low productivity
Livestock mortality and morbidity Human morbidity and mortality	Human morbidity and mortality Livestock morbidity and mortality Number of schooldays missed due to illness Frequency of diseases affecting kale	Types and frequency of human diseases Causes of human morbidity

(continued on next page)

TABLE 6.2 (continued)
Village-Level Community-Based Agroecosystem Health Indicators, Kiambu District, Kenya, June 1998

Attribute	Mahindi	Kiawamagira	Gitangu
Markets	Location of nearest market Quantity of farm produce going to market	Variety of goods available in the shopping center	Variety of goods in the market
Savings/ wealth	Types of houses Number of livestock per homestead	Number of cattle per homestead	Increasing or decreasing needs in homesteads
Knowledge	Types of skills	Farming techniques Behavior of youth and children	Knowledge of current affairs Frequency of extension visits
Infrastructure	Distance to primary schools Status of access road		Status of schools, medical facilities and roads

described for the LUU-level indicators. The indicator on rainfall was based on data to be obtained from the meteorological department based on weather stations closest to each of the study sites. The indicator on physical fertility of soils was based on data to be obtained from the Ministry of Agriculture and Kenya Agricultural Research Institute's soil classification databases.

6.3.3 Indicator Measurement and Refinement

6.3.3.1 Community Driven

The groups assigned the duty of carrying out empirical measurements of community-driven indicators met three to four times in a span of 1 month between August and September 1998 to discuss their methods and findings. A final report of the findings was presented in a village workshop with the research team present in October 1998. Table 6.6 shows a summary of the reports by village.

In some cases, participants did not give a measurement. The initial statement was either vague or too circumspect. Further probing by facilitators failed to yield any clarification. The following illustrates a common trend during the sessions:

Group leader: Indicators for market availability were distance to nearest market and quantity of produce going to the market. We found that these were good indicators.

Facilitator: Could you say whether the markets are near or far and whether the produce taken to the market is a lot or just a little?

Group leader: I cannot answer that question.

Gikabu-na-buti	Thiririka	Githima
Number and location of outlets for produce	Number and location of outlets for produce	Demand versus supply of produce (price) Access to markets
Permanent houses Number of tea bushes	Number of children not going to school due to lack of school fees	Tea bushes Coffee bushes
Farming techniques Number of schools and attendance Attendance to hospitals	Frequency of extension meetings	Farming techniques Number of people with technical skills
Types of buildings		Quality of access road Type of buildings

In most cases when no statements were given for an indicator, there were indications that a discussion had taken place during the group meeting and a consensus reached on how to make the report. These were most likely situations in which a consensus on what to report was not reached, participants were unable to carry out the measurements, or cultural factors inhibited public debate. There were difficulties in recording actual morbidity and mortality data (with respect to both humans and livestock). Where information on the number of deaths was given, the target population and the time period covered was not supplied. Most communities preferred not to quantify morbidity and mortality. There were indications that participants in all villages had difficulty dealing with quantities and numerical measurements. Participants preferred, and were able to analyze, nominal data (e.g., very high, high, low, and very low).

For a number of attributes, participants dropped some of the indicators and selected new ones. The reasons given were that some indicators were difficult to measure or the information gathered was not easy to interpret or not useful at all. It was difficult to elucidate the processes followed since the research team was not present during the group discussions.

6.3.3.2 Researcher Proposed

Table 6.7 shows the means and standard errors of the quantitative, researcher-proposed LUU-level indicators. In 7.1% (16/225) of the LUUs, all the adults (non-school-going persons 18 years and older) were involved in off-farm activities. However, the average number of people dependent (for employment) on 1 acre of crop fields was 22.69 ± 1.55 persons, with an average monthly per capita income of 1,339.77 ± 179.43 shillings. In contrast, the average monthly wage was 6,537.11 ± 1,179.47 shillings.

TABLE 6.3
Attributes, Categories, and Number of Researcher-Proposed Indicators of Health and Sustainability of the Kiambu Agroecosystem

	Attribute	Category	LUU[a]	SS[b]
Biophysical	Biophysical efficiency	Allocative	10	10
		Technical	5	5
	Environmental quality	Chemical pollution	1	2
		Rainfall	nil	1
		Tree cover	nil	1
	Pests, diseases, and health	Animal	1	1
		Crops	1	1
		Demographics	nil	2
		Health and nutrition	5	1
		Human	3	3
	Soil fertility	Chemical	1	nil
		Physical	nil	1
	Water	Availability	2	2
		Quality	1	1
Economic	Capital	Credit	nil	1
		Investments	8	7
	Farm efficiency	Inputs	3	nil
		Outputs	3	3
		Profitability	1	1
	Income	Amount	2	2
		Nonfarm	1	1
		Savings	1	1
	Infrastructure	Accessibility	1	nil
		Condition	nil	nil
Social	Aspirations	Achievements	nil	nil
		General goals	nil	nil
		Satisfaction	1	nil
	Attitudes	Education	2	nil
		Health	nil	nil
		Professions	nil	nil
		Wealth	nil	nil
		Work	nil	nil
	Equity	Control	1	nil
		Ownership	1	nil
		Roles	nil	nil
		Social values	nil	nil
	Knowledge and information	Formal	1	1
		Informal	1	1
		Innovativeness	nil	nil
		Sources	2	3
		Technology	nil	nil

TABLE 6.3 (continued)
Attributes, Categories, and Number of Researcher-Proposed Indicators of Health and Sustainability of the Kiambu Agroecosystem

Attribute	Category	LUU[a]	SS[b]
Linkages	Contacts	1	1
	Familial ties	2	2
	Outmigration	nil	nil
Organization	Family structure	nil	nil
	Leadership	nil	nil
	Organizations	1	1
	Reciprocity	1	nil
	Social control	nil	nil
Preferences	Farm enterprises	2	nil
	Food	1	nil
	Leisure	nil	nil
	Occupations	1	1
Values	Behavioral	nil	nil
	Wealth related	nil	nil
	Well-being	nil	nil

[a] Number of land-use unit-level indicators
[b] Number of study-site-level indicators

LUUs with no cattle comprised 27.1% (61/225) of the total. There was an average of 1.36 ± 0.11 cattle per acre. The average acreage of land used for agriculture per LUU was 2.86 ± 0.39, comprising 104.0% of the total land owned. An average of 13.0% of the area used for farming in a LUU was rented. Among the indicator crops, the proportion of land under maize was the largest (0.32 ± 0.02), followed by land under beans (0.21 ± 0.02). Although acreage under kale was small relative to other indicator crops, yield in kilograms per acre was the highest, followed by that of potatoes. The average milk yield was 2.92 ± 0.24 kg per cow per day.

The average number of sick days per person per month was 1.92 ± 0.21, with only 0.07 ± 0.01 hospital visits per person per year and 0.03 ± 0.00 hospitalizations per person per annum, on average. However, the average annual expenditure on health per LUU was 13,276.03 ± 3,659.65 shillings. Of the LUUs, 140 (62%) did not have children less than 5 years of age. Of the 85 that had children in this age group, 32.9% (28/85) did not have CHC cards for any of these children.

Most (64%) of the LUUs did not experience morbidity in livestock, but most (78%) reported experiencing crop pests and diseases (Table 6.8). The soil fertility score was low for most (91%) of the LUUs. Most (92%) of the LUUs obtained their water from a source less than 1 km away. Most (74%) owned bank accounts, but only a few had coffee (8%) or tea (16%) production. Most (69%) had at least one contact with an extension worker in a year. Most (60%) reported that farm productivity was satisfactory.

Of the variability in the land-use-level, researcher-proposed indicators, 70% was accounted for by the first 34 dimensions of the MCA (Table 6.9). The first dimension

TABLE 6.4
Researcher-Proposed Land-Use Unit (LUU)-Level Indicators of Health and Sustainability

Classification			Indicator	Name
Biophysical	Biophysical efficiency	Allocative	1. Off-farm employment rate[a]	OffFarm
			2. Heads of cattle/available labor	CattleLabor
			3. Available labor per acre	AcreLabor
			4. Heads of cattle per acre	CattleAcre
			5. Proportion of land under indicator crops[b]	Land
			6. Proportion of farmland rented	PropRent
			7. Napier production	Napier
		Technical	8. Per acre yield of indicator crops[b]	Yield
			9. Milk yield/cow/day	MilkYield
	Environmental quality	Chemical	10. Expenditure on agrochemicals	AgChemExp
	Pests, diseases, and health	Animal	11. Morbidity in cattle	CattleMorbidity
		Crops	12. Occurrence of plant diseases	PlantDcz
		Health and nutrition	13. Recorded child health clinic visits per child	HealthVisits
			14. Proportion of children with health cards	HealthCards
			15. Recorded vaccination events/child	Vaccinations
			16. Annual expenditure on health	HealthExp
			17. Weight-age ratio of children	WeightAge
		Human	18. Hospital visits/person/month	HospVisits
			19. Hospitalizations/person/year	Hospitalized
			20. Sick days/person/month	SickDays
	Soil fertility	Chemical	21. Soil fertility score	Soil
	Water	Availability	22. Distance to water source	WaterDist
			23. Monthly expenditure on water	WtrExpend
		Quality	24. Coliform counts	Coliforms
Economic	Capital	Investments	25. Coffee production	Coffee
			26. Tea production	Tea
			27. Proportion of farmland owned[c]	PropOwn
			28. Heads of cattle	Cattle

TABLE 6.4 (continued)
Researcher-Proposed Land-Use Unit (LUU)-Level Indicators of Health and Sustainability

Classification			Indicator	Name
			29. Number of sheep and goats	Shoats
			31. Total acreage of farmland[d]	AreaAgric
			32. Proportion of indicator resources owned	Resources
	Farm efficiency	Inputs	33. Income/inputs for nonfood crops	CBRCashCrop
			34. Income/inputs for food crops	CBRFoodCrop
			35. Income/inputs for livestock	CBRLivst
		Outputs	36. Income per acre of nonfood crop	IncPACC
			37. Income per acre of food crop	IncPAFC
		Profitability	38. Profitability[e]	Profitability
	Income	Amount	39. Per capita farm income[f]	PerCapt
			40. Average wage per employed person	Wage
		Nonfarm	30. Proportion of income that is nonfarm	NonFarm
		Savings	41. Ownership of a bank account	BankAccount
	Infrastructure	Accessibility	42. Infrastructure within walking distance	Access
Social	Aspirations	Satisfaction	43. Farm productivity score	prodScore
	Attitudes	Education	44. School dropout rate[g]	DropOuts
		Education	45. Annual expenditure on education	EdnExpend
	Equity	Control	46. Female control of indicator resources[h]	GenderCtrl
		Ownership	47. Female ownership of indicator resources[i]	GenderOwn
	Knowledge and information	Formal	48. Proportion of adults with postprimary education	Education
		Informal	49. Frequency of clan meetings	ClanMeet
		Sources	50. Extension contact	Extension
	Linkages	Contacts	51. Frequency of visits to friends	VisitsF
		Familial ties	52. Frequency of visits to relatives	VisitsR
			53. Proportion of family[j] living outside village	OutRel

(continued on next page)

TABLE 6.4 (continued)
Researcher-Proposed Land-Use Unit (LUU)-Level Indicators of Health and Sustainability

Classification			Indicator	Name
	Organization	Organizations	54. Membership in community-based organizations	Membership
		Reciprocity	55. Frequency of exchanges[k]	Reciprocity
	Preferences	Farm enterprises	56. Prop of common foods produced in LUU	FoodPdcC
			57. Prop of traditional foods produced in LUU	FoodPdcT
		Food	58. Proportion of traditional foods eaten	FoodEatT

[a] Number of adults with off-farm employment over total number of persons in LUU
[b] Maize, beans, potatoes, kale
[c] Acreage for which a title deed exists over total acreage used by members of the LUU
[d] Total acreage used by LUU members for farming and dwelling
[e] Total cash income minus total cash expenditure on farm enterprises
[f] Total cash income from farm enterprises over the total number of persons in LUU
[g] Number of non-school-going persons below 19 years of age over total number of persons in LUU
[h] Proportion of indicator resources controlled by females
[i] Proportion of indicator resources owned by males
[j] Nuclear family members only
[k] Exchange of material and service gifts (*itega*) among LUU

accounted for 6.1% of the total variation in the data, the second 5.5%, the third 4.0%, the fourth 3.6%, and the fifth and sixth 3.1% and 3.0%, respectively, totaling 25.3%. Each of dimensions 7 to 34 accounted for between 2.7% and 1.1% of the total variation, amounting to 45.4%. The principle inertias ranged from 0.101 for dimension 1 to 0.020 for dimension 34, indicating that the dimensions accounted for significant variability (correlations between the indicators and the scores of these dimensions) in the data ($p < .05$).

Indicators most correlated with the scores of the 34 dimensions are shown in Table 6.9. The scores of first and fourth dimensions were most correlated with heads of cattle ($r^2 = 0.53$). In dimension 1, the factor loadings (coordinates) decreased with increasing numbers of cattle per LUU (H [high] = −0.93; L [low] = −0.15, and N [none] = −1.07). Dimension 4 was a contrast between LUUs with few cattle (L = −0.44) and those with none (N = 0.88). In this dimension, LUUs with more cattle had the least inertia (H = −0.08). Dimension 2 was a contrast of LUUs in which nonfarm income was reported against those that did not respond. The third dimension was a contrast between LUUs that produced beans and those that did not.

All categories of the 66 indicators were fairly well represented by the 34 dimensions, except the number of sheep and goats (= L with a quality of 0.46). Categories

TABLE 6.5
Researcher-Proposed Study-Site-Level Indicators of Health and Sustainability for the Kiambu Agroecosystem, Kenya, 1998

Classification			Indicator	Name
Biophysical	Biophysical efficiency	Allocative	1.Proportion of LUU with napier	Napier
			2. Proportion land under indicator crops/LUU	
			3. Cattle per available labor	CattleLabor
			4. Proportion of LUU renting land	Landrent
			5. Cattle per acre	CattleAcre
			6. Available labor per acre	AcreLabor
			7. Leasing out land	LandLease
		Technical	8. Yield/acre of indicator crops	
			10. Milk yield per cow/day	MilkYield
	Environmental degradation	Chemical pollution	13. Average expenditure on agrochemicals	AgChemExp
			14. Proportion LUU using agrochemicals	PropAgChem
		Rainfall	Mean monthly rainfall	Rainfall
		Tree cover	15. Proportion of LUU with woodlots	Woodlots
	Pests, diseases, and health	Animal	16. Proportion LUU with animal diseases	AnimDcz
		Crops	17. Proportion LUU with crop pests and disease	PlantDcz
		Demographics	Persons per LUU	LUUSize
			LUU per square kilometer	Density
		Human health	18. Proportion of LUU with health cards	HealthCards
		Human diseases	19. Proportion of LUU with hospital visits	HospVisits
			20. Proportion of LUU with hospitalizations	Hospitalized
			21. Sick days/person/month	Sickdays
	Soil fertility	Physical	Soil classification	Soil
	Water	Availability	22. Proportion LUU far from water source	WaterDist
			23. Average expenditure on water	WtrExpend
		Quality	24. Coliform counts/LUU	Coliforms
Economic	Capital	Credit	25. Proportion of LUU that took credit	Credit

(continued on next page)

TABLE 6.5 (continued)
Researcher-Proposed Study-Site-Level Indicators of Health and Sustainability for the Kiambu Agroecosystem, Kenya, 1998

Classification			Indicator	Name
		Investments	26. Heads of cattle per LUU	Cattle
			27. Sheep and goats per LUU	Shoats
			29. Acreage of farmland/LUU	AreaAgric
			30. Proportion of LUU with coffee production	Coffee
			31. Proportion of LUU with tea production	Tea
			32. Indicator-resource ownership/LUU	Resources
	Farm efficiency	Outputs	33. Income from cash crop	IncPACC
		Profitability	34. Average profitability	Profitability
	Income	Amount	35. Employment rate	Employ
			36. Income per person	PerCapt
		Nonfarm	28. Proportion LUU with nonfarm income	NonFarm
		Savings	37. Proportion of LUU with bank accounts	BankAccount
Social	Knowledge and information	Formal	38. Postprimary education per LUU	Education
		Informal	39. Grandparents living with grandchildren	GrandChild
		Sources	40. Proportion LUU with extension contacts	extension
			41. Proportion LUU with frequent visits to friends	VisitsF
			42. Distance to Nairobi	NbiDist
	Linkages	Familial ties	43. Nuclear family outside village	OutRel
			44. Proportion LUU with frequent visits to relatives	VisitsR
	Organization	Organizations	45. Average membership in CBOs	Membership

CBO, community-based organization; LUU, land-use unit

with the lowest mass included expenditure on agrochemicals = none, school dropouts = present, head of cattle per available labor = missing, income per acre of cash crop = high, hospitalizations per person per year = high, income per acre of cash crop = low, recorded vaccination events/child = high, per capita income = missing, coffee production = present, distance to water source = far, and soil fertility score = high in order of increasing mass. Those with the highest mass included proportion of indicator common foods eaten = low, distance to water source = close, coffee production = absent, soil fertility score = low, napier production = absent, and tea production = absent.

The first dimension has a score that is correlated with measures of allocative efficiency of cattle production (heads of cattle [0.65], head of cattle/available labor [0.56], and heads of cattle per acre [0.54]). The score of the second dimension is most correlated with the proportion of LUU income that is nonfarm ($r^2 = 0.30$) and with the school dropout rate ($r^2 = 0.29$). The highest correlations with the score of the third dimension are with measures of allocative efficiency of food crop production (bean yield per acre [0.36], proportion of land under beans [0.35], maize yield per acre [0.33], proportion of land under maize [0.32], potatoes yield per acre [0.25], proportion of land under potatoes [0.22]).

Table 6.10 shows the principle inertia of the six dimensions accounting for 75.9% of the variation in researcher-proposed, SSL (village-level) indicators of health and sustainability, the indicators most correlated with the score of each dimension and the coordinates for their categories along these dimensions. The first and second dimensions accounted for over 16% of the variation each, while each dimension from the third to the sixth accounted for between 12% and 8%. The principle inertias ranged from 0.19 for dimension 1 to 0.08 for dimension 6. Only the first three dimensions represented significant average correlations between the indicators and the scores ($p < .1$).

Five categories had a quality less than 0.6 (distance to Nairobi = L [0.40], all categories of nuclear family outside village [0.46], and all categories of occurrence of animal diseases [0.50]). The category with the lowest mass was soil classification = H (0.003). The ones with the highest (0.014) mass were coffee production = A, nuclear family outside village = L, proportion of farms using agrochemicals = L, proportion of LUUs with bank accounts = L.

6.3.4 Comparison of Indicator Suites

Six of the attribute classifications were common to both suites of indicators: (1) equity, (2) environmental quality, (3) soil fertility, (4) pest and disease dynamics, (5) infrastructure, and (6) knowledge. However, the focus was on different categories of indicators within each of the attribute classes, resulting in differences in the indicators chosen. For example, communities focused mostly on productivity and physical characteristics in the soil fertility attribute, while the researcher-proposed suite focused on chemical fertility and physical classification of the soils. The choice of indicators within the same category of an attribute differed between the two suites. Among the indicators common to both suites were distance to water source, frequency of hospital visits, number of livestock, availability of extension services, accessibility of infrastructure, morbidity and mortality, quantities of yields, and presence or absence of various farm enterprises. The use of livestock numbers and cash crops as indicators of capital, wealth, or savings was common to both suites. An important difference between the two suites was the presence of value-based measures such as "proper hygiene," "good behavior," "good variety," and "good habits" in the community-based suite. In addition, many of the indicators in this suite were mostly in ordinal scale. Researcher-proposed indicators were mostly numeric, non-value-based measures generally on the continuous scale.

TABLE 6.6
Summary of Indicator Evaluations Carried Out by Communities in the Intensive Study Site, Kiambu District, Kenya, August–September 1998

Attribute	Mahindi	Kiawamagira	Gitangu
Lifestyle	No statement given[a,b]	Some people do not farm. No novel farming techniques. Christian values have modified culture. No female circumcision. Christian weddings/marriages predominant. Traditional food types and cooking methods are disappearing. Most houses made of timber and/or iron sheets.	No clear assessment.[a]
Social organization	Lack of unity caused failure of most plans. Planning was inadequate.	Very good. Meetings held regularly. Considerable progress in CAP (Community Action Plans) implementation. Number of groups few, indicating lack of unity.	Persistence of water problem indicates ineffective leadership.
Equity	No statement given.	Meetings balanced in terms of gender and age. "Women do most of the farmwork and household chores, while men keep most of the farm income."	Gender relations fair. Women share in leadership positions. Income from farm is owned by both.
Quality of environment	Air quality good. Streams are clean. There is need to increase use of latrines, improve garbage disposal, and boil drinking water.	Waterborne diseases (diarrhea) very common. Smell and effluent from slaughterhouses pollute. "Jiggers, indicating insufficient water for domestic use, afflict many children."	Lack of water lowers the hygiene standards. Many varieties of chemicals used. There is need for proper disposal of these materials.
Soil fertility	Only 3 farms have fertile soil. Fertility improved by use of manure.	Weeds[c] that indicate soil fertility not found. Signs of erosion in every farm.	Average soil fertility. Nearly all farms use manure. Erosion evident in some farms.
Farm productivity	The yields are too low. No granaries at all.	More dairy cattle than 4 years ago. Maize, beans, and cabbages purchased.	Productivity low because of poor management and lack of water. Most produce consumed on farm.

Gikabu-na-buti	Thiririka	Githima
Timber and/or iron sheet houses majority. Few permanent buildings (school, offices).	"We used to eat cold and tasteless meals. Now we have hot meals and even meat."	Diets have less maize and beans.
Nursery school and church projects implemented. Other statements circumspect.	Primary and nursery school projects. Meetings frequent. Good attendance.	Community participation high. Projects implemented: primary and secondary schools, water, cattle dip.
Statement circumspect.		Very few women attend meetings. Respect between age groups eroded. Young people have no land. Dressing code different.
Statement circumspect.	Pit latrines and cowsheds too close to wells in most homesteads. Disposal of agrochemical. Wells near vegetable plots.	Too much dust. Latrines poorly constructed in most homesteads. Some homesteads still using river water.
Statement circumspect.	Poor soils in three-quarters of farms.	Poor: crops less green (more yellowish).
Many people take produce to market. Few food items purchased.	Milk yield averages 2 kg/cow/ day. Kale yields 100 kg/ fortnight.	Low yields due to poor farming techniques.

(contnued on next page)

TABLE 6.6 (continued)
Summary of Indicator Evaluations Carried Out by Communities in the Intensive Study Site, Kiambu District, Kenya, August–September 1998

Attribute	Mahindi	Kiawamagira	Gitangu
Pests and diseases	No statement given.	Death[d] rate high. Morbidity high.[d] Malaria, typhoid, and alcoholism main causes.	Morbidity high.[d] Causes were coughing, common cold, tuberculosis, and malaria.
Markets	No statement given.	Market is well supplied with goods.	Sufficient variety of goods in market.
Savings	No statement given.	Currently purchasing most food items.	Most farms have livestock cattle.
Knowledge	No statement given.	No new farming techniques.	Knowledge of current affairs is high. Extension meetings regular. A number of farms currently using novel farming techniques.
Infrastructure	Access road in very poor condition.	Access road in very poor condition.	School, access road in fair condition.

[a] Statements were not clear and did not refer to previously selected indicators.
[b] Participants did not give findings for this indicator. Follow-up questions resulted in noncommittal answers or decline to answer.
[c] Because these weeds were absent, researchers were unable to establish their identity.
[d] Participants did not want to provide numbers.
[e] No indication of the period considered. Participants did not wish to provide details.
[f] Refers not only to constipation but also to heartwater.
[g] A skin disease in sheep. Exact etiology being confirmed. Most likely sheep keds.

6.4 DISCUSSION

6.4.1 Comparison of Indicator Suites

With the researcher-proposed indicators focusing mostly on numeric, non-value-based measures, it was difficult to find suitable measures in the social domain and less so in the economic domain. In contrast, community-based indicators were more strongly value based, focusing mostly on a social-economic interpretation of the underlying biophysical phenomena. The community-based suite contained many indicators that would be suitable for many of the attributes in the social domains of the researcher-proposed suite. The two suites therefore provided complementary information on the health and sustainability of the agroecosystem. That this was the case is further supported by the fact that communities requested to be provided with

Gikabu-na-buti	Thiririka	Githima
Very high morbidity. Causes: malaria, fever, pneumonia, and diarrhea. High livestock mortality last year. Last potato crop affected by bacterial wilt.	Morbidity high during the cold weather. Twelve people have died.[e] Causes were asthma, tuberculosis, and flu. Livestock diseases: Ndigana[f] and konji.[g]	Morbidity high, mostly due to malaria, coughing, tuberculosis, and malnutrition. Need to increase vegetables in our diet.
Outlets for vegetables, milk, tea, potatoes adequate.	Most produce rots in the farm.	Poor access to markets. No control of prices. Spoilage of produce (milk, tea).
One section of village has good houses and a lot of tea crop.	No statement given.	Coffee or tea crops in many homesteads. Few or no livestock in many homesteads.
No statement given.		Few people with technical skills.
Roads and buildings in fair condition.	Roads condition fair. School condition poor.	The access road is in poor condition.

a report of the findings from the researcher-proposed indicator measurement. These reports were followed by intense community discussions.

That the two suites measure very similar agroecosystem attributes is probably a reflection of the fact that the researcher-proposed suite was based on community goals and felt needs. This supports the view that indicators based on community goals and felt needs are likely to be more managerially useful, considering that communities are the primary managers of agroecosystem. Because communities often lack the capacity to develop and measure non-value-based indicators, while researchers and policymakers lack the knowledge and mandate to make value-based judgments, it seems that decision support systems for such integrated and adaptive approaches as sustainability and agroecosystem health should include both components to provide a balanced assessment.

TABLE 6.7
Summary Statistics for Quantitative Land-Use Unit (LUU)-Level Researcher-Proposed Indicators Measured in 225 LUU in 12 Villages of Kiambu District, Kenya

	Attribute	Category	Description	Mean	SEM[a]	Missing	Zero
Biophysical	Biophysical efficiency	Allocative	Available labor per acre	22.69	1.55	0	0
			Head of cattle/available labor	1.17	0.12	16	56
			Heads of cattle per acre	1.36	0.11	0	61
			Off-farm employment rate	0.16	0.01	0	107
			Proportion of farmland rented	0.13	0.02	0	169
			Proportion of land under beans	0.21	0.02	0	92
			Proportion of land under kale	0.08	0.01	0	115
			Proportion of land under maize	0.32	0.02	0	45
			Proportion of land under potatoes	0.14	0.01	0	85
		Technical	Bean yield per acre	77.19	14.97	0	160
			Kale yield per acre	9,390.44	3,015.86	0	164
			Maize yield per acre	765.77	267.72	0	128
			Milk yield/cow/day	2.92	0.24	0	107
			Potato yield per acre	1,541.02	364.00	0	136
	Envtal quality	Chemical pollution	Expenditure on agrochemicals	6,111.11	1,553.58	24	11
	Pests, diseases, and health	Health and nutrition	Annual expenditure on health	13,276.03	3,659.65	24	32
			Average weight-age ratio	1.13	0.13	173	0
			Proportion of children with health cards	0.62	0.05	140	28
			Recorded child health clinic visits per child	6.46	0.44	171	0
			Recorded vaccination events/child	7.07	0.34	171	0
		Human diseases	Hospital visits/person/month	0.07	0.01	0	146
			Hospitalizations/person/year	0.03	0.00	0	182
			Sick days/person/month	1.92	0.21	0	105

	Water	Availability	Monthly expenditure on water	162.37	29.55	48	124
		Quality	Coliform counts	272.21	101.18	71	0
Economic	Capital	Investments	Heads of cattle	2.27	0.16	0	61
			Number of sheep and goats	1.70	0.22	0	137
			Proportion of farmland owned	1.04	0.07	0	32
			Proportion indicator resources owned	0.66	0.01	0	0
			Total acreage of farmland	2.86	0.39	0	0
	Farm efficiency	Inputs	Income/inputs for food crops	2.91	0.83	124	0
			Income/inputs for livestock	2.38	0.91	146	0
			Income/inputs for nonfood crops	1.53	0.56	179	0
		Outputs	Income per acre of food crop	13,638.96	2,711.02	125	0
			Income per acre of nonfood crop	59,490.72	11,476.52	190	0
		Profitability	Profitability	11,875.43	3,960.63	56	0
	Income	Amount	Average wage	6,537.11	1,179.47	150	0
			Per capita farm income	1,339.77	179.43	18	0
		Nonfarm	Proportion of income that is nonfarm	0.34	0.03	73	69
	Infrastructure	Accessibility	Infrastructure within walking distance	0.58	0.02	0	0
Social	Attitudes	Education	Annual expenditure on education	24,908.80		0	81
	Equity	Control	Female control of indicator resources	0.51	0.02	0	24
		Ownership	Female ownership of resources	0.33	0.02	0	48
	Knowledge	Formal	Proportion with postprimary education	0.35	0.03	0	103
	Linkages	Familial ties	Proportion family outside village	0.32	0.02	0	72
	Organization	Organizations	Membership in community-based organizations	1.63	0.06	45	0
	Preferences	Farm enterprises	Proportion common foods produced	0.69	0.01	0	0
		Farm enterprises	Proportion of traditional foods produced	0.44	0.02	0	26
		Food	Proportion of traditional foods eaten	0.78	0.01	0	0

[a] Standard error of the mean.

TABLE 6.8
Summary Statistics for Qualitative Land-Use Unit (LUU)-Level Researcher-Proposed Indicators of Agroecosystem Health Measured in 225 LUU in 12 Villages of Kiambu District, Kenya

	Attribute	Category	Description	Categories		
Biophysical	Biophysical efficiency	Allocative	Napier production	A = 0.86	P = 0.14	
	Pests, diseases, and health	Animal	Morbidity in cattle	A = 0.64	P = 0.36	
		Crops	Occurrence of plant diseases	A = 0.22	P = 0.78	
	Soil fertility	Chemical fertility	Soil fertility score	H = 0.09	L = 0.91	
	Water	Availability	Distance to water source	H = 0.08	L = 0.92	
Economic	Capital	Investments	Coffee production	A = 0.92	P = 0.08	
		Investments	Tea production	A = 0.84	P = 0.16	
	Income	Savings	Ownership of a bank account	A = 0.26	P = 0.74	
Social	Aspirations	Satisfaction	Farm productivity score	H = 0.12	L = 0.28	S = 0.60
	Attitudes	Education	School dropout rate	A = 0.65	E = 0.29	P = 0.06
	Knowledge and information	Informal	Frequency of clan meetings	H = 0.22	L = 0.31	N = 0.47
		Sources	Extension contact	A = 0.31	P = 0.69	
	Linkages	Contacts	Frequency of visits to friends	H = 0.15	L = 0.51	N = 0.35
		Familial ties	Frequency of visits to relatives	H = 0.15	L = 0.56	N = 0.28
	Organization	Reciprocity	Frequency of exchanges	H = 0.12	L = 0.48	N = 0.40

A, absent; E, nonresponse; H, high; L, low; N, none; P, present; S, satisfactory

6.4.2 Indicator Measurement and Refinement

A major constraint to the community-based indicator measurement was the cultural inhibitions to discuss certain issues in public. Although selected as indicators of health and diseases by most communities, human morbidity and mortality, for example, were not discussed in public, and data were not supplied. This diminishes the value of the indicators and the overall assessment as a decision support tool. A more positive aspect was that the data for most of the indicators were on an ordinal scale, making interpretation and overall assessment easier. In some instances, however, the data gathered were mostly anecdotal and therefore of limited value. More

TABLE 6.9
Dimensions Accounting for 70% of the Variation Among 226 Land-Use Units Based on Multiple Correspondence Analysis of Researcher-Proposed Indicators of Agroecosystem Health and Sustainability, Kiambu, Kenya, 1999

Dim	Inertia[a] (%[b])	Indicator[c] (r^2)[d]	Categories (Coordinates[e])			
1	0.101 (6.05)	Heads of cattle (0.53)	H (−0.93)	L (−0.15)	N (1.07)	
2	0.092 (11.56)	Proportion of income that is nonfarm (0.30)	E (−0.71)	H (0.26)	L (0.76)	N (0.12)
3	0.069 (15.60)	Bean yield per acre (0.36)	E (−0.46)	H (−0.49)	L (−0.58)	N (0.71)
4	0.061 (19.15)	Heads of cattle (0.31)	H (−0.08)	L (−0.44)	N (0.88)	
5	0.054 (22.25)	Kale yield per acre (0.34)	E (0.37)	H (−0.46)	L (−1.15)	N (0.37)
6	0.052 (25.24)	Heads of cattle per acre (0.18)	H (−0.53)	L (0.47)	N (0.03)	
7	0.046 (27.89)	Female ownership of resources (0.23)	H (0.58)	L (−0.11)	N (−0.71)	
8	0.043 (30.34)	Frequency of visits to relatives (0.15)	H (−0.91)	L (0.22)	N (0.05)	
9	0.041 (32.70)	Annual expenditure on health (0.35)	E (0.39)	H (0.60)	L (−0.29)	N (−1.07)
10	0.040 (34.98)	Bean yield per acre (0.18)	E (−0.44)	H (−0.11)	L (0.89)	N (0.01)
11	0.038 (37.10)	Proportion of farmland rented (0.20)	H (0.74)	L (−0.97)	N (0.09)	
12	0.037 (39.18)	Frequency of visits to friends (0.21)	H (1.07)	L (−0.29)	N (−0.03)	
13	0.036 (41.17)	Head of cattle/available labor (0.15)	E (−1.17)	H (−0.20)	L (0.32)	N (0.11)
14	0.035 (43.11)	Farm productivity score (0.17)	H (−0.34)	L (−0.57)	S (0.33)	
15	0.033 (44.95)	Frequency of clan meetings (0.14)	H (−0.61)	L (0.45)	N (−0.01)	
16	0.032 (46.70)	Female ownership of resources (0.18)	H (−0.07)	L (0.40)	N (−0.71)	
17	0.031 (48.41)	Kale yield per acre (0.38)	E (−1.12)	H (−0.37)	L (0.24)	N (0.46)
18	0.030 (50.07)	Coliform counts (0.11)	E (0.30)	H (0.18)	L (−0.44)	
19	0.030 (51.70)	Coliform counts (0.15)	E (0.29)	H (−0.54)	L (0.25)	
20	0.029 (53.26)	Hospitalizations/person/year (0.20)	H (−0.80)	L (1.13)	N (−0.09)	
21	0.027 (54.74)	Off-farm employment rate (0.13)	H (−0.58)	L (0.42)	N (−0.01)	
22	0.027 (56.19)	Female ownership of resources (0.11)	H (−0.28)	L (−0.08)	N (0.61)	

(continued on next page)

TABLE 6.9
Dimensions Accounting for 70% of the Variation Among 226 Land-Use Units Based on Multiple Correspondence Analysis of Researcher-Proposed Indicators of Agroecosystem Health and Sustainability, Kiambu, Kenya, 1999

Dim	Inertia[a] (%[b])	Indicator[c] (r^2)[d]	Categories (Coordinates[e])			
23	0.026 (57.62)	Maize yield per acre (0.14)	E (−0.41)	H (0.05)	L (0.58)	N (0.02)
24	0.026 (59.02)	Distance to water source (0.13)	C (0.11)	F (−1.20)		
25	0.024 (60.32)	Frequency of visits to friends (0.24)	H (0.50)	L (0.31)	N (−0.67)	
26	0.023 (61.56)	Monthly expenditure on water (0.14)	E (−0.33)	H (−0.64)	L (0.73)	N (0.10)
27	0.023 (62.79)	Milk yield/cow/day (0.12)	H (−0.51)	L (0.45)	N (−0.02)	
28	0.022 (63.98)	Heads of cattle (0.10)	H (0.44)	L (−0.30)	N (0.15)	
29	0.022 (65.15)	Proportion of adults with postprimary education (0.11)	H (−0.17)	L (0.53)	N (−0.23)	
30	0.021 (66.28)	Recorded vaccination events/child (0.11)	E (0.04)	H (−1.03)	L (0.34)	
31	0.021 (67.39)	Expenditure on agrochemicals (0.13)	E (−0.79)	H (−0.01)	L (0.08)	N (1.13)
32	0.020 (68.47)	Monthly expenditure on water (0.10)	E (−0.24)	H (0.02)	L (−0.71)	N (0.24)
33	0.020 (69.53)	Sick days/person/ month (0.16)	H (0.39)	L (−0.63)	N (0.15)	
34	0.020 (70.58)	Prop income that is nonfarm (0.08)	E (−0.29)	H (0.39)	L (−0.25)	N (0.24)

A, absent; C, close; Dim, dimension; E, missing; F, far; H, high; L, low; N, none; P, present

[a] Principal inertias (also the average squared correlations; Greenacre, 1993)

[b] Cumulative percentage of total inertia accounted for by the dimensions based on adjusted principal inertias (Greenacre, 1993)

[c] The indicator most highly correlated with the scores derived from optimal scale values

[d] Correlation coefficient

[e] Also the optimal scale values for the categories

reliable and useful assessments were obtained when communities visited each other and made comparative assessments.

The main difficulty with the researcher-based suite was that the initial list—especially the LUU level—consisted of long lists of indicators, resulting in even longer questionnaires that were difficult to administer. The refined LUU level consisted of 34 indicators, which is still too long a list for follow-up empirical measurements given that most indicators were designed to be measured using questionnaires and

TABLE 6.10
Dimensions Accounting for Over 75% of the Variation Among 12 Study Sites (Villages) and Based on a Multiple Correspondence Analysis of Study-Site-Level, Researcher-Proposed Indicators of Agroecosystem Health and Sustainability

Dim	Inertia[a] (%[b])	Indicator[c] (r^2)[d]	Categories (Coordinates[e])		
1	0.188 (18.03)	Indicator-resource ownership/LUU (0.72)	H (1.00)	L (−0.72)	
2	0.175 (34.76)	Distance to Nairobi (0.72)	H (−1.41)	L (0.14)	M (0.73)
3	0.146 (48.58)	Coffee production (0.58)	A (−0.44)	P (1.32)	
4	0.115 (59.32)	Sheep and goats per LUU (0.48)	H (0.82)	L (−0.59)	
5	0.095 (68.13)	Coliform counts/LUU (0.48)	H (−0.82)	L (0.59)	
6	0.084 (75.85)	Kale yield/acre (0.51)	H (−0.85)	L (0.60)	

A, absent; C, close; Dim, dimension; E, missing; F, far; H, high; L, low; LUU, land-use unit; N, none; P, present

[a] Principal inertias (also the average squared correlations; Greenacre, 1993)

[b] Cumulative percentage of total inertia accounted for by the dimensions based on adjusted principal inertias (Greenacre, 1993)

[c] The indicator most highly correlated with the scores derived from optimal scale values

[d] Correlation coefficient

[e] Also the optimal scale values for the categories

participatory tools. Since the process is adaptive, the lists can be further refined as more understanding of the system is gained. In addition, the larger data sets serve as reference points to help interpret results from the more refined suites.

The use of correspondence analysis had three main advantages. First, the relationships between various indicators were complex and did not often fit the multivariate normal distributional requirements of most analytical approaches for continuous data. More crucial though was the presence of nonresponses and zero values that carried special meaning but would be discarded in most analytical methods for continuous data. Finally, interpretation of the indicators would only be meaningful relative to a set of cutoff or threshold values, necessitating a transformation of continuous variables to an ordinal scale. Correspondence analysis provided a means for summarizing and graphically presenting the data in a way that enabled identification of important trends and for reducing the dimensions of the indicator suite.

6.4.3 Practicality and Application

The process used in this research project is predicated on two main assumptions. The first is that there exists a community with a set of common goals and values within the targeted study sites. The second is that there exists sufficient capacity for collective action to enable negotiation and compromise if the goals and values are competing or conflicting and for the development of a community-based monitoring system.

The combination of participatory methods and soft system methodology provides a means through which goals and values can be stated and negotiated. The use of the health language is crucial in helping the communities to build a conceptual framework of the indicator selection process and the measurement and assessment processes. Presenting the agroecosystem as an individual whose health status is unknown, indicators are defined as those measures that need to be taken to make a statement on the health status on such an individual.

The results of this process demonstrate that agroecosystem health and sustainability can be used as part of a community's decision support system. Communities created action plans and revised and implemented them based on the information derived from indicators. The process of indicator selection and measurement in itself appeared to enhance collective action, while sensitizing communities to the underlying biophysical and socioeconomic processes that had an impact on the health and sustainability of their agroecosystem. Some practical improvements to the process include the use of geographical information systems to enhance the quality and cost-effectiveness of indicator measurement, development of self-reporting systems, and automation of some reporting and feedback processes, which would increase the cost-effectiveness of the overall process.

REFERENCES

Ayres, R.U. (1996). Statistical measures of unsustainability. *Ecological Economics* 16: 239–255.

Boyle, M., Kay, J.J., and Pond B., 2000. Monitoring in support of policy: An adaptive ecosystem approach. In *Encyclopaedia of Global Environmental Change*. Munn, R., ed. New York: Wiley, pp. 116–137.

Cairns, J., Jr., McCormick, P., and Niedrelehner, B.R. (1993). A proposed framework for developing indicators of ecosystem health. *Hydrobiologia* 263: 1–44.

Canadian Council of Ministers of the Environment (CCME). (1996). *A Framework for Developing Ecosystem Health Goals, Objectives and Indicators: Tools for Ecosystem-Based Management.* Manitoba Statutory Publications, Winnipeg, Manitoba, Canada, p. 24.

Casley, D.J., and Lury, D.A. (1982). *Monitoring and Evaluation of Agricultural and Rural Development Projects.* John Hopkins University Press, London. 145 pp.

Dumanski, J. (1994). *A Framework for Evaluation of Sustainable Land Management (FESLM).* Agriculture and Agri-Food Canada, Ottawa.

Gallopin, G.C. (1994a). Agroecosystem health: A guiding concept for agricultural research? In *Agroecosystem Health with Special Reference to the Consultative Group for International Research (CGIAR): Proceedings of an International Workshop.* Nielsen, N.O., ed. Guelph, Ontario, Canada: Agroecosystem Health Project, Faculty of Environmental Sciences, University of Guelph, pp. 94–97.

Gosselin, P., Belanger, D., Bibeault, J.F., and Webster, A. (1991). *Feasibility Study on the Development of Indicators for a Sustainable Society.* Health and Welfare Canada, Quebec.

Harrington, L.W. (1992). Measuring sustainability: Issues and alternatives. In *Let Farmers Judge: Experiences in Assessing the Sustainability of Agriculture.* Hiemstra, W., Reijntjes, C., and van der Werf, E., eds. London: Intermediate Technology, pp. 26–41.

Izac, A.-M.N., and Swift, M.J. (1994). On agricultural sustainability and its measurement in small-scale farming in sub-Saharan Africa. *Ecological Economics* 11: 105–125.

Kay, J.J. (1993). On the nature of ecological integrity: some closing comments. In *Ecological Integrity and the Management of Ecosystems.* Woodley, S., Kay, J., and Francis, G., eds. Delray, FL: St. Lucie Press, pp. 210–212.

Lightfoot, C., and Noble, R.P. (1993). A participatory experiment in sustainable agriculture. *Journal of Farming Systems Research Extension* 4(1): 11–34.

National Research Council. (1993). *Sustainable Agriculture and the Environment in the Humid Tropics.* National Academy Press, Washington, DC, p. 702.

Rapport, D.J. (1992). Evolution of indicators of ecosystem health. In *Ecological Indicators.* McKenzie, D.H., Hyatt, D.E., and McDonald, V.J., eds. London: Elsevier Applied Science, pp. 121–134.

Rapport, D.J., and Regier, H.A. (1980). An ecological approach to environmental information. *Ambio* 9: 22–27.

Rapport, D.J., Regier, H.A., and Hutchinson, T.C. (1985). Ecosystem behavior under stress. *American Naturalist* 125: 617–640.

Schaeffer, D.J., Herricks, E.E., and Kerster, H.W. (1988). Ecosystem health: I. Measuring ecosystem health. *Environmental Management* 12: 445–455.

Smit, B., Waltner-Toews, D., Rapport, D., Wall, E., Wicher, G., Gwyn, E., Wandel, J. (1998). *Agroecosystem Health: Analysis and Assessment.* Agroecosystem Health Project, University of Guelph, Ontario, Canada, p. 114.

Winograd, M. (1994). *Environmental Indicators for Latin America and the Caribbean: Toward Land Use Sustainability.* IICA/GTZ Project, World Resources Institute (WRI), Washington, DC. 83 pp.

7 An Assessment of Health and Sustainability of a Smallholder-Dominated Tropical Highlands Ecosystem

7.1 INTRODUCTION

In an agroecosystem health and sustainability assessment, indicators should be analyzed in two ways: (1) as measures of overall health at a point in time and (2) as predictors of its long-term sustainability (Costanza et al., 1998) and health. Assessing the health status of an agroecosystem involves comparing and contrasting a series of indicator measurements against a set of cutoff and threshold values (Canadian Council of Ministers of the Environment, 1996) based on the goals and objectives of the agroecosystem.

A suite of indicators would in most cases contain several dozen variables. A method of summarizing and presenting such data must preserve its holistic and multidimensional nature while providing meaningful, quantitative, and easily understood criteria for evaluating agroecosystem health. One approach is to combine indicators into indices such as total factor productivity (Ehui and Spencer, 1993), ecosystem health index (Costanza, 1992), and agricultural sustainability index (Nambiar et al., 2001). A fatal disadvantage of this approach is that indices place weights on different indicators without providing a rational basis for their (the weights') choice. Another disadvantage is that these indices would eventually require some form of decomposition to provide managerially useful information—a decomposition that more often involves a reassessment of the initial suite of indicators used to compute the index—and back to the initial problem of how to summarize information from indicators. Less unencumbered by the latter, but still crippled by the weighting problem, are approaches such as ecological footprint (Wackernagel and Rees, 1997) and the method proposed by Afgan et al. (2000) based on decision support systems and the general indices method.

A systems approach to evaluating indicator data requires an understanding of how agroecosystem goals and values are seen to relate to each other and to the various social, biophysical, and economic phenomena that underlie the indicators. Understanding the phenomena that data from indicators portray, however, requires a systemic approach for two reasons. First, indicators are representations of complex phenomena within a self-organizing, goal-seeking complex system. While these phenomena are controlled by feedback mechanisms, they present mainly as stochastic processes with a high level of unpredictability and further complicated by differential effects across scales and time spans. Second, agroecosystems often have multiple, sometimes-competing goals, and the objective of the system is goal optimization rather than maximization. Furthermore, the process of goal optimization involves a series of trade-offs and balances within the system and between the system and the external environment. To obtain managerially useful information from indicators, there needs to be a systemically generated conceptual framework that delineates the expectations from system goals in terms of both their impact and the inputs required to achieve them. The health status of the system can then be obtained by assessing the implication of various indicator values (outcomes) with regard to generic health attributes such as integrity, adaptability, resilience, efficiency, efficacy, effectiveness, vigor, and productivity.

Predictions on the long-term sustainability and health of the systems rely on an analysis of spatial and temporal trends of the indicators (Rapport and Regier, 1980; Odum, 1985; Rapport et al., 1985). Interpretations of these trends require a systems approach as well. A potentially useful approach is to use dynamic models such as pulse processes to assess generic system attributes of the system given the trends portrayed by the indicator data. Using contrasts between point measurements and targets or thresholds, scenarios at different spatial and time spans can be re-created and evaluated relative to a set of goals. Trends in indicators can be modeled as trends in pulses within such models. Graphical techniques—especially plots in multidimensional Euclidean space—provide intuitive tools for summarizing and presenting data in forms that aid identification of such trends. Simple correspondence analysis (SCA) and multiple correspondence analysis (MCA) are especially attractive tools for exploring trends in indicators (Gitau et al., 2000) by enabling the categorization of data based on predefined cutoffs and thresholds while not requiring any distributional assumptions.

This chapter describes how community participation, cognitive maps, and correspondence analysis were used to evaluate indicator data. The objective was to generate managerially useful information that can be used to guide practical human activity in the Kiambu agroecosystem.

7.2 PROCESS AND METHODS

7.2.1 Spatial and Temporal Trends in the Indicators

The objective was to determine, based on indicator measurements, what were the most significant differences among the villages and in each village along the time line of the project. In addition, the response of the holons to the project as an external "stress" was compared across the six intensive study sites (ISSs) and along the project time line. The extensive study sites were included in some of the analyses as controls, to increase statistical power, and in the calculation of cutoffs, ranges, and thresholds for indicators.

Researcher-proposed indicators were used in MCA to generate visual and descriptive summaries of the trends in the indicator data. Two empirical measurements were carried out on the same study sites and the same land-use units (LUUs) within each study site, first in January–March 1999 and then in January–March 2000. The methods used for measuring the indicators are described in Chapter 6. Data were managed using a relational database (Microsoft Access) and analyzed using SAS (SAS Institute Inc., SAS Campus Drive, Cary, NC 27513).

Simple correspondence analysis (PROC CORRESP) was used to explore the spatial trends in the indicator data. The analysis was based on a cross tabulation of the study sites by each of the researcher-proposed LUU-level indicators. The analysis and interpretation were based on the methods described by J. M. Greenacre (1993) and M. Greenacre and Blasius (1994). Study site points that were close together were considered as representing similarities along the respective dimensions, while those that were further apart were considered as indicating differences along the plotted dimensions.

For the temporal analysis, indicator measurements for the second round of measurements were offered as supplementary points in MCA (Benzecri, 1992) of the 1999 data. The correlation between the coordinates of the main points and the supplementary points was used to determine the presence of significant deviation of inertia between the two measurements. Dimensions with the smaller Pearson's coefficients were considered to have important temporal trends. The statistical significance of these trends was assessed by testing the null hypothesis $(1 - r) = 0$, where r is the correlation coefficient. Points that were further away from the main diagonal of a plot of the main coordinates against the supplementary ones were considered to represent a deviation in the inertia along that dimension over time. The further away the point, the more significant the deviation. The significance of these deviations was assessed by comparing the proportion of LUUs in the specific category during the first indicator measurement to those in the same category in the second round of measurements.

7.2.2 Evaluation of Goals, Expectations, and Achievements

The objective was to explain, in a systemic way, the values, patterns, and trends in indicators based on the perceived progress in community goals. Were the goals reasonable given the available resources? Were the expected benefits reasonable given the underlying social, economic, and biophysical processes? Given the changes implemented, what would be the reasonable expectations over the short-, medium-, and long-term time spans?

Progress toward community goals was evaluated using participatory methods. Participants were asked to rank progress as negative, stagnant, slight, moderate, or a lot. The ranking tools used were as described in Chapter 2. Evaluation of progress was carried out in 1-day participatory workshops in January 1999 and in January 2000. The changes in the system perceived to be driving this progression were also recorded.

Community expectations for each of these goals were assessed based on pulse process models of their cognitive maps. The expected primary outputs were those changes in system attributes that would be expected to be the direct result of implementing the action plans and strategies. Expected secondary outputs were those changes in the system attributes resulting from the cascading effects of the implemented action plans. Changes and patterns in indicators were evaluated based on these expectations to decide whether these were met and to evaluate the suitability of the indicators, the validity of some community assertions, and the impact of community goals on the agroecosystem.

The implications of the spatial and temporal trends in the indicators in terms of generic system health attributes such as productivity, stability, integrity, adaptability, resilience, efficiency, efficacy, effectiveness, vigor, and equitability were assessed based on the communities cognitive maps. Discrete dynamic models based on the cognitive maps were used in this assessment. Details to the models are provided in Chapter 4.

7.3 RESULTS

7.3.1 Spatial Trends in the Researcher-Proposed Indicators

Figure 7.1 is a scatterplot of the first two dimension of an SCA of village- against researcher-proposed LUU-level indicators measured in 1999. Together, they accounted for 36.5% of the total inertia. Most of the villages were clustered together in the upper right quadrant, except Kameria, which was in the lower left, and Kihenjo, Githima, and Gitwe, all of which were in the lower right. Dimension 1 has high negative weights on income/acre of cash crops and income/inputs for cash crops and higher positive weights on kale yield and production of traditional foods. Factor levels with high absolute loads along dimension 2 were water expenditure, distance to water source, and coliform counts. Figure 7.2 shows a similar scatterplot using the January 2000 indicator data. The distribution pattern of column points was similar to that from the 1999 data. The characteristics of the first two dimensions, account-

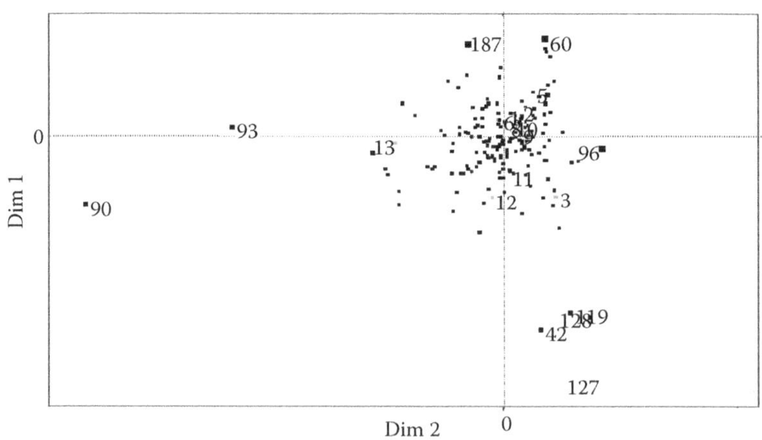

FIGURE 7.1 Scatterplot of dimension 1 against dimension 2 in a simple correspondence analysis of village against researcher-proposed land-use unit (LUU)-level indicators measured in 1999. See CD for color image and key.

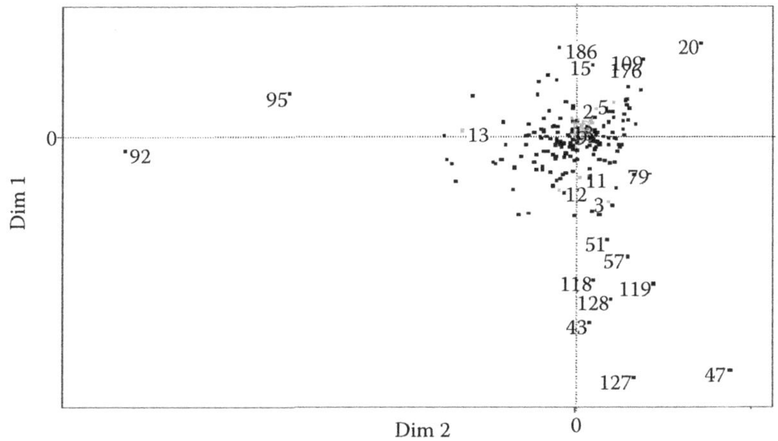

FIGURE 7.2 Scatterplot of dimension 1 against dimension 2 in a simple correspondence analysis of village against researcher-proposed land-use unit (LUU)-level indicators measured in January 2000. See CD for color image and key.

ing for 39.8% of the total inertia, changed only slightly, with the first dimension relating strongly to production characteristics of the LUU and the second relating to water availability and quality.

7.3.2 Evaluation of Temporal Trends in the Researcher-Proposed Indicators

The dimensions with the lowest correlation coefficients were 15 ($r = 0.72$), 19 ($r = 0.74$), 23 ($r = 0.74$), and 3 ($r = 0.75$). Figure 7.3 shows the change in inertia of categories along dimension 15 between the January 1999 (DIM13R1) and the January 2000 (DIM13R2) indicator measurements. Categories 39 (inputs/income for livestock = H [high]), 40 (inputs/income for livestock = L [low]), 58 (proportion traditional foods = H), 81 (hospitalizations = L), 83 (hospital visits = H), 101 (maize yield = H), 122 (per capita income = E), 130 (potatoes yield = H), 170 (average wage = H), and 171 (wage = L) showed the most change along this dimension.

Categories 16 (beans yield = H), 80 (hospitalizations = H), 84 (hospital visits = L), 95 (kale yield = H), 122 (per capita income = E), 130 (potatoes yield = H), 133 (production score = H), and 136 (profitability = E) showed the most change in inertia along dimension 19 (Figure 7.4). Among those that showed the most change along dimension 23 were categories 16 (beans yield = H), 17 (beans yield = H), 81 (hospitalizations = L), 96 (kale yield = L), 102 (maize yield = L), and 130 (potatoes yield = H) (Figure 7.5). Along dimension 3, the most change in inertia was by categories 3 (available labor/acre = H), 35 (cost/inputs of food crops = E), 35 (cost/inputs of food crops = L), 80 (hospitalizations = H), 89 (income/acre of food crops = E), 89 (income/acre of food crops = L), 96 (kale yield = L), 122 (per capita income = E), and 136 (profitability = E) (Figure 7.6).

Figure 7.7 shows the location of categories along dimensions 15 and 19 based on an MCA of the 1999 measurements with the January 2000 measurements as

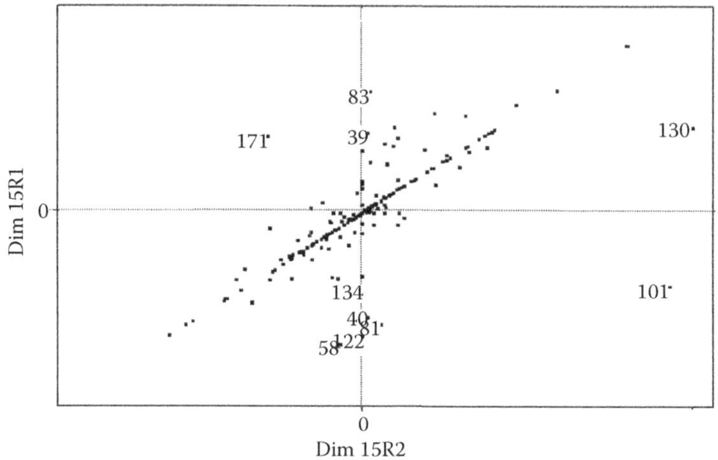

Key

Numbered points = Categories with most change in inertia: 39 (Inputs/income for livestock = H), 40 (Inputs/income for livestock = L), 58 (Prop traditional foods = H), 81 (Hospitalizations = L), 83 (Hospital visits = H), 101 (Maize yield = H), 122 (Per capita income = E), 130 (Potatoes yield = H), 134 (Productivity score), 170 (Average wage = H), and 171 (Wage = L)

FIGURE 7.3 Change in inertia of land-use unit (LUU)-level indicator categories along multiple correspondence analysis (MCA) dimension 15 between January 1999 (DIM15R1) and January 2000 (DIM15R2).

supplementary points. Among categories with a shift toward the center of gravity were 1 (inputs/income for livestock = H), 2 (inputs/income for livestock = L), 3 (proportion traditional foods = H), 4 (hospitalizations = L), 6 (maize yield = H), 8 (potatoes yield = H), 9 (productivity score), 12 (beans yield = H), 13 (hospitalizations = H), 95 (kale yield = H), 133 (production score = H), and 136 (profitability = E). Among those with a shift away from the center were 5 (hospital visits = H), 7 (per capita income = E), and 11 (wage = L). Category 10 (average wage = H) had a sign inversion.

The distribution of categories along dimensions 3 and 23 is shown in Figure 7.8. Among the categories that moved toward the center were 3 (hospitalizations = L), 4 (kale yield = L), 5 (maize yield = L), 8 (cost/inputs of food crops = E), 9 (cost/inputs of food crops = L), 12 (income/acre of food crops = L), 13 (per capita income = E), and 14 (profitability = E). Category 6 (potatoes yield = H) moved away from the center while categories 1 (beans yield = H) and 2 (beans yield = L) had sign inversion.

Table 7.1 is a summary of the test of significance of the trends in proportions of LUUs having characteristics identified through MCA as showing important temporal trends based on the two indicator measurements. Trends in income/inputs for food crops, income/acre of food crops, profitability, average wage, and per capita farm income were related to improvements in the response rate. Changes in the technical biophysical efficiency were a significant ($p < .001$ for each) decrease in the number of LUUs classified as having high yields of indicator crops (beans, maize, kales, and potatoes). In terms of economic farm efficiency, there was a significant increase in

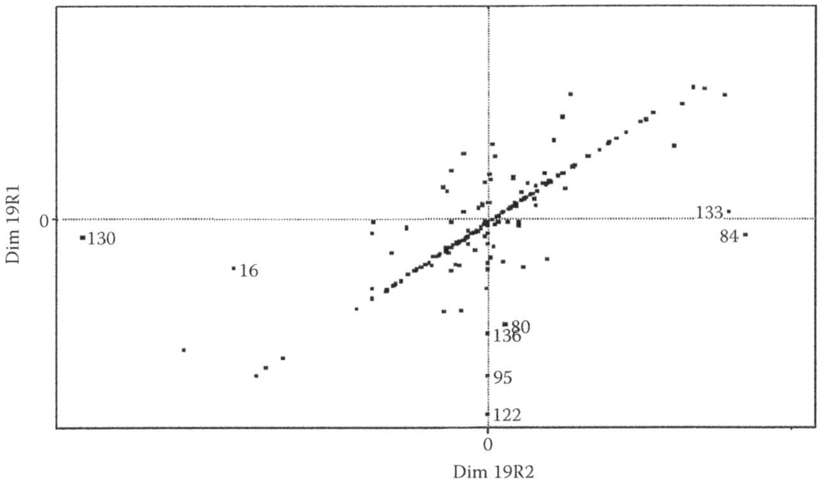

Key

Numbered points = Categories with most inertia change: 16 (Beans yield = H), 80 (Hospitalizations = H), 84 (Hospital visits = L), 95 (Kale yield = H), 122 (Per capita income = E), 130 (Potatoes yield = H), 133 (Production score = H) and 136 (Profitability = E)

FIGURE 7.4 Change in inertia of categories along multiple correspondence analysis (MCA) dimension 19 between January 1999 (DIM19R1) and the January 2000 (DIM19R2) measurements of researcher-proposed land-use unit (LUU)-level indicators.

the number of LUUs with high income/inputs for livestock. Changes in pest, disease, and health dynamics were marked by a significant increase in the number of hospital visits/person/month ($p < .001$) and the number of hospitalizations per person per year ($p < .001$). Significantly more LUUs reported an increase in the proportion of indicator traditional foods eaten ($p < .001$).

7.3.3 Evaluation of Goals, Expectations, and Achievements

Table 7.2 shows progress toward community goals in the six ISSs as of January 1999 and January 2000. All villages had improved contact with extension staff, which was attributed to the improvement of many of the agriculture-related goals, such as crop productivity and reduction in crop pests and diseases. Similarly, there was reported to be an improvement in security, with reduction in crime rates in nearly all villages where this was considered a problem. In Mahindi and Kiawamagira, there were initial attempts by the communities to improve the access roads that resulted in only slight improvements. In Githima, addition of classrooms in the existing school was reported to result in only slight improvements in literacy and school attendance. An initial attempt to obtain water from a pipeline passing near the village had only slight-to-moderate success as only a small section of the village was receiving water by January 1999. By January 2000, the situation had improved markedly. In addition, the access roads in Githima were graded, and this was reported as moderate improvement.

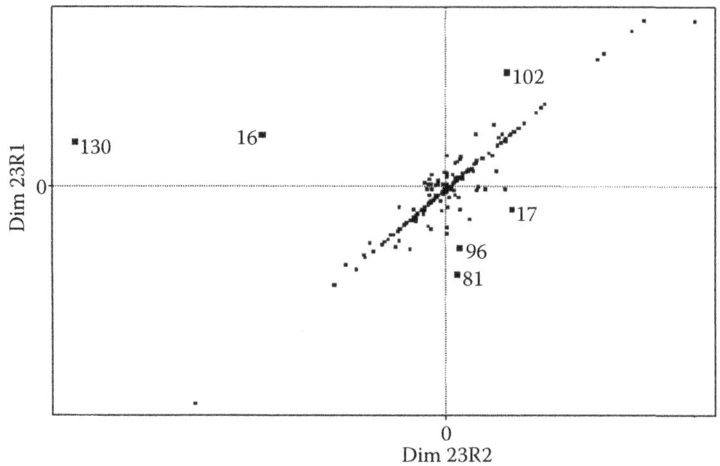

Key

Numbered points = Categories with highest change in inertia: 16 (Beans yield = H), 17 (Beans yield = H), 81 (Hospitalizations = L), 96 (Kale yield = L), 102 (Maize yield = L) and 130 (Potatoes yield = H)

FIGURE 7.5 Change in inertia of categories along multiple correspondence analysis (MCA) dimension 23 between January 1999 (DIM23R1) and January 2000 (DIM23R2) measurements of researcher-proposed land-use unit (LUU)-level indicators.

Improved health care in Kiawamagira was reported to be due to improved access to a privately owned health facility near the area. Communities reported that the activities resulting in increased contact with extension staff and the improved security could be maintained. Similarly, supply of water to most households in Githima village could be sustained over the long term, as could the road maintenance. Communities in Mahindi and Kiawamagira carry out routine maintenance of access roads, but the condition of the road was ranked as only a slight improvement. Table 7.3 shows the changes in system attributes resulting from these activities and the expected primary and secondary outputs based on a pulse process model of the communities' cognitive maps.

In Githima village, the expected outputs included improvements in coffee, tea, and dairy production, resulting in increased farm productivity and household incomes as well as an improvement in knowledge, literacy, and employment opportunities, resulting in reduction in the number of people dependent on farmland for their livelihoods (Table 7.3). The community foresees deterioration in soil productivity as a possible outcome of this process. In Gitangu village, the expected outputs were an improvement in the farming techniques, resulting in improved poultry, dairy, and crop production, resulting in improved income and human health. In Kiawamagira, the primary expectations were an improvement in human health due to improved health care and increasing non-farm employment through small-scale enterprises, building of rental houses, and access to jobs outside the village. Improved access road was expected to result in enhanced dairy and flower production and increased access to off-farm jobs in Mahindi. In Gikabu, the expected outputs were an improvement

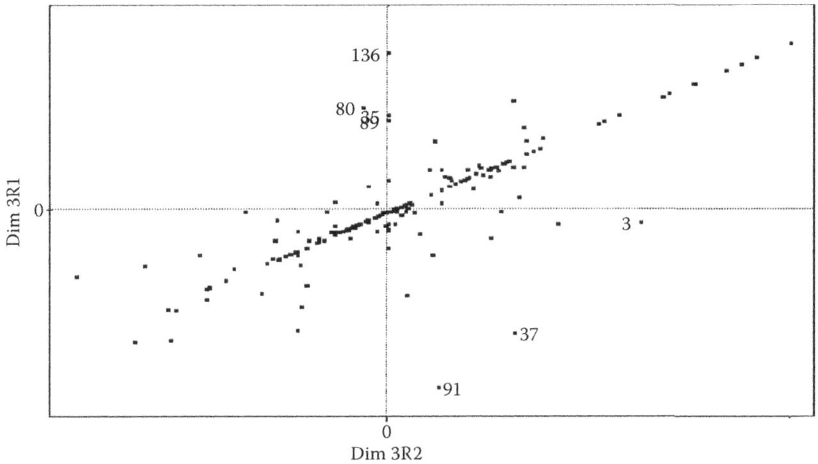

Key

Numbered points = Categories with highest inertia change: 3 (Available labor/acre = H), 35 (Cost/inputs of food crops = E), 35 (Cost/inputs of food crops = L), 80 (Hospitalizations = H), 89 (Income/acre of food crops = E), 89 (Income/acre of food crops = L), 96 (Kale yield = L), 122 (Per capita income = E) and 136 (Profitability = E)

FIGURE 7.6 Change in inertia of categories of the researcher-proposed land-use unit (LUU)-level indicators along multiple correspondence analysis (MCA) dimension 3 between January 1999 (DIM3R1) and January 2000 (DIM3R2).

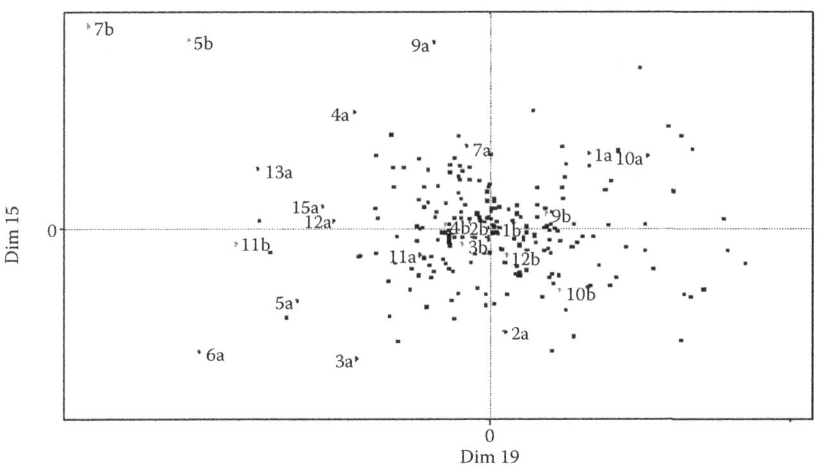

FIGURE 7.7 Scatterplot of dimension 15 against dimension 19 showing change in inertia of categories between January 1999 and January 2000. See CD for color image and key.

in the production of tea and other crops due to improved farming techniques, eventually leading to improved nutrition and incomes. Farm labor shortage and increasing vermin population were seen as potential negative outcomes. In Thiririka, the

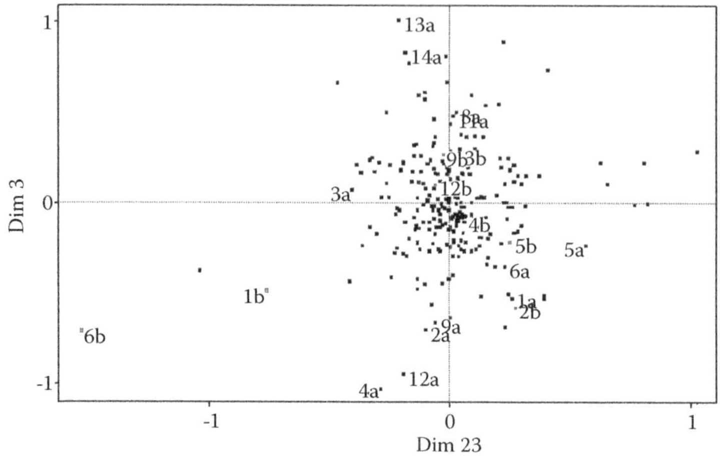

FIGURE 7.8 Scatterplot of dimension 3 against dimension 23 showing change in inertia of categories between January 1999 and January 2000. See CD for color image and key.

expected outputs were an improvement in human health and in incomes. However, increases in crop diseases were foreseen, with these eventually leading to negative impacts in terms of agrochemical use.

7.4 DISCUSSION

With only two rounds of measurements over a 2-year period, it is difficult to assess the agroecosystem based on the trends in the indicators. Further measurements would be required to provide a more valid assessment of health and sustainability. However, the methods used in this study demonstrate an approach that may be useful in summarizing and presenting indicator data. The advantages of correspondence analysis are twofold: (1) the incorporation of targets and thresholds in the process of categorizing the indicators, thus providing an intuitive interpretation, and (2) projection of data from the initial and subsequent measurements into a multidimensional space, with the distribution of points easily interpretable in terms of the χ^2 distribution.

7.4.1 Spatial and Temporal Trends in the Indicators

Simple correspondence analysis grouped villages based on two main criteria: the crop production characteristic and water availability patterns. This is in agreement with the data from the participatory process, in which water was identified as an important constraint and cash crop production as an important source of household income and a determinant of land use in the district.

Spatial trends were confounded by the changes in response rates for many indicators and possible interviewer bias. The response rate was increased for many of the indicators between the first and the second measurements. This is more likely due to the feedback provided to the communities subsequent to the first

TABLE 7.1
Land-Use Unit (LUU) Indicators Identified Through Multiple Correspondence Analysis as Showing Important Temporal Trends Between January 1999 and January 2000

Attribute	Category	Indicator	Level	1999 (%)	2000 (%)	p^a
Biophysical efficiency	Technical	Bean yield per acre	H	32 (14.22)	5 (2.22)	< .001
			L	33 (14.67)	55 (24.44)	= .01
		Kale yield per acre	H	30 (13.33)	0 (0.00)	< .001
			L	31 (13.78)	120 (53.33)	< .001
		Maize yield per acre	H	47 (20.89)	6 (2.67)	< .001
			L	50 (22.22)	75 (33.33)	= .01
		Potato yield per acre	H	43 (19.11)	2 (0.89)	< .001
Farm efficiency	Inputs	Income/inputs for food crops	E	124 (55.11)	0 (0.00)	< .001
			L	51 (22.67)	93 (41.33)	< .001
		Income/inputs for livestock	H	39 (17.33)	101 (44.89)	< .001
			L	40 (17.78)	79 (35.11)	< .001
	Outputs	Income per acre of food crop	E	125 (55.56)	0 (0.00)	< .001
			L	51 (22.67)	169 (75.11)	< .001
	Profitability	Profitability	E	56 (24.89)	0 (0.00)	< .001
Income	Amount	Average wage	H	33 (14.67)	99 (44.00)	< .001
			L	42 (18.67)	46 (20.44)	= .63
		Per capita farm income	E	18 (8.00)	0 (0.00)	< .001
Pests, diseases, and health	Human diseases	Hospital visits/person/month	H	33 (14.67)	151 (67.11)	< .001
		Hospitalizations/person/year	H	17 (7.56)	75 (33.33)	< .001
			L	26 (11.56)	7 (3.11)	< .001
Preferences	Food	Proportion of traditional foods eaten	H	37 (16.44)	145 (64.44)	< .001

H, high; L, low; E, missing
Probability for Ho: Proportion of LUU in 1999 = Proportion of LUU in 2000 against Ha: Proportion of LUU in 1999 ≥ = Proportion of LUU in 2000

measurement and the recognition by them that this was useful information. Because of this, many of the farmers who had not been able to provide estimates on yields of various indicator crops were able to do so in the subsequent round of measurements, while those who had been unwilling to provide income-related information were willing to do so. With further measurements, it would be possible to assess the impact of these on the accuracy of the measurements. In addition, other methods of collecting data on the indicators should be explored to minimize the cost and the interviewer biases inherent in the methods used in this study.

The possibility of an overall reduction in the technical biophysical efficiency of most LUUs between the first and the second measurements is indicated by the

TABLE 7.2
Progress in Community Goal Achievements in the Intensive Study Sites as Assessed in January 1999 and January 2000

	January 1999						January 2000					
Goals	Gikabu	Gitangu	Githima	Kiawamagira	Mahindi	Thiririka	Gikabu	Gitangu	Githima	Kiawamagira	Mahindi	Thiririka
Agrochemical use						1						1
Crop diseases	0					1	1					1
Distance to schools				0						0		
Food shortage	1						1					
Fuel shortage			0	0					0	0		
Grabbing of public land						2						2
Ignorance			0						0			
Illiteracy			1						1			
Inadequate extension	2	3				2	2	3				2
Inadequate security	3	3	3	1		3	3	3	3	2		3
Lack of artificial insemination			0	0					0	0		
Lack of market and shopping center		0						0				
Lack of market for tea	1						2					
Lack of nursery schools					0						0	
Lack of organization and unity						1						1
Lack of secondary school and polytechnic		0						0				
Low crop productivity				1						1		
Low dairy productivity				1						1		
Low-quality seeds						0						0
No telephones				0						0		
Pests and diseases		0						0				
Poor access road(s)		0	2	1	1			0	2	1	1	
Poor health care system	0		0	1	0	0	0		0	1	0	0
Poor human health	0	0	0			1	0	0	0			1
Poor leadership						1						1
Poor-quality feeds		1						1				
Soil infertility and erosion		1				1		1				1
Substance abuse	0						0					
Unemployment	0	0			0		0	0			0	
Water not accessible	0	0	1	0	0		0	1	3	0	0	
Water not potable					0						0	

0, no progress; 1, slight progress; 2, moderate progress; 3, much progress significant decline in the number of the LUUs classified as having high yield per acre of indicator cash crops. It is more likely that improved reporting by farmers (reduction in recall bias) would result in an increase in the estimated yields rather than a decrease. Further support for this is the significant increase in the number of LUUs reported as consuming a high proportion of indicator traditional foods, many of which are utilized during periods of reduced food availability.

7.4.2 Evaluation of Goals, Expectations, and Achievements

While there was progress in some of the goals, it is clear that communities did not follow the priority ranking that they had made in their revised action plans. The reason for this does not appear to be a change in priorities but probably is a reflection of the difficulty in implementing some of the plans. In all but two of the villages, the activities undertaken were those requiring the least investment in terms of money and time/labor. The formation of vigilante groups and organization of extension workshops are a good example of this. In all villages, there were attempts to implement the first item in the action plans, with varying degrees of success among them. It is difficult to compare the villages based on this because each problem situation was unique, requiring unique approaches and resources to fulfill. It is however remarkable that the communities that had most success in implementing their action plans (based on the number and success of the projects) were those that were older settlements. The exception to this was Mahindi village, which was unique in the sense that all the leaders in the village were young people (mostly below 25 years of age). The reason for this seems to be that most of the adults in this village were either very old or very young. The middle-aged people lived away from the village (mostly in Nairobi), where they had formal employment.

Many of the goals seemed to be confluent with the communities' expectation of ecosystem health and sustainability, except in Githima and Thiririka, where use of agrochemicals had both positive and negative impacts, therefore requiring optimization. Based on the cognitive maps, however, some community expectations were far beyond what could be achieved. An example is the expectation, in Githima village, that building classrooms would result in increased literacy levels in the same time span as it takes improved soil fertility to result in increased household incomes.

TABLE 7.3
Expected Primary and Secondary Outputs, Based on a Pulse Process Model, Resulting from Interventions in the Intensive Study Sites as of January 1999

Attribute	Githima			Gitangu			Kiawamagira			Mahindi			Gikabu			Thiririka		
	C	1	2	C	1	2	C	1	2	C	1	2	C	1	2	C	1	2
Use of agrochemicals																+	+	−
Artificial insemination services		+	+						+									
Crop diseases														+	+	+	−	−
Coffee production		+	+															
Crop productivity							+		+					+				
Pests and diseases					+	+												
Dairy production		+			+		+				+	+						
Extension services				⊕	+	+	⊕	+	+				⊕	+	+	⊕	+	+
Feed quality				⊕	+	+												
Flower production											+	+						
Fodder								+										
Food crop production					+	+												
Food shortage													+					
Fuel availability			+															
Public land preservation																⊕		
Health care							⊕	+	+									
Human health						+		+								+	+	
Horticulture																		+
Hygiene		+																
Knowledge		+	+															
Illiteracy	+	+	+															
Income		+			+	+									+		+	+
Small-scale enterprises								+										

Kale production									+									
Labor export		+	+					+	+		+	+						
Labor shortage															–			
Availability of farmland			+															
Quality of leadership																⊕		
Manure						+												
Migration												–						
Nutrition														+				
Private veterinarians								+				+						
Farm productivity		+																
Poultry production					+													
Rental houses								+										
Roads	⊕	+	+				⊕	+	+	⊕	+	+						
School committee			–															
Schools	⊕	+																
Security	⊕	+		⊕	+		⊕	+					⊕	+				
Soil			–	+				+	+							+		
Tea estates															+			
Tea market													⊕	+	+			
Tea centers			+															
Tea production		+	+											+	+			
Farming techniques			+		+	+												
Unemployment												+						
Unity																⊕		
Vermin															–			
Water availability	⊕	+				+												
Water project			+															

C, changes; 1, expected primary output; 2, expected secondary output; ⊕, changes that communities assumed they would be able to maintain; +, positive impact; – negative impact

REFERENCES

Afgan, N.H., Carvalho, M.G., and Hovanov, N.V. (2000). Energy system assessment with sustainability indicators. *Energy Policy* 28: 603–612.

Benzecri, J.P. (1992). *Correspondence Analysis Handbook.* Marcel Dekker Inc., New York, New York. 125 pgs.

Canadian Council of Ministers of the Environment. (1996). *A Framework for Developing Ecosystem Health Goals, Objectives and Indicators: Tools for Ecosystem-Based Management.* Manitoba Statutory Publications, Winnipeg, Manitoba, Canada, p. 24.

Costanza, R. (1992). Towards an operational definition of health. In *Ecosystem Health: New Goals for Environmental Management.* Costanza, R., Norton, B.G., and Haskell, B.D., eds. Washington DC: Island Press, pp. 239–256.

Costanza, R., Mageau, M., Norton, B., and Patten, B.C. (1998). Predictors of ecosystem health. In *Ecosystem Health.* Rapport, D.J., Costanza, R., Epstein, P.R., Gaudet, C., and Levins, R., eds. London: Batsford, pp. 240–250.

Ehui, S., and Spencer, D.S.C. (1993). Measuring the sustainability and economic viability of tropical farming systems: A model from sub-Saharan Africa. *Agricultural Economics* 4(9): 279–296.

Gitau, T., McDermott, J.J., Waltner-Toews, D., and McDermott, B.M. (2000). Use of correspondence analysis in agroecosystem health assessments. In *Proceedings of the Ninth Symposium of the International Society for Veterinary Epidemiology and Economics, August 6–11, 2000.* Colorado: ISVEE, Breckenridge, pp. 501–502.

Greenacre, J.M. (1993). *Correspondence Analysis in Practice.* Academic Press, Orlando, FL, p. 195.

Greenacre, M., and Blasius, J. (1994). *Correspondence Analysis in the Social Sciences: Recent Developments and Application.* Academic Press, Orlando, FL, p. 370.

Nambiar, K.K.M., Gupta, A.P., Fu, O., Li, S. (2001). Biophysical, chemical and socio-economic indicators for assessing agricultural sustainability in the Chinese coastal zone. *Agriculture, Ecosystems and Environment* 87: 209–214.

Odum, E.P. (1985). Trends in stressed ecosystems. *Bioscience* 35: 419–422.

Rapport, D.J., and Regier, H.A. (1980). An ecological approach to environmental information. *Ambio* 9: 22–27.

Rapport, D.J., Regier, H.A., and Hutchinson, T.C. (1985). Ecosystem behavior under stress. *American Naturalist* 125: 617–640.

Wackernagel, M., and Rees, W. (1997). Perceptual and structural barriers to investing in natural capital: Economics from an ecological footprint perspective. *Ecological Economics* 20: 3–24.

8 General Discussion

8.1 SYNOPSIS

The general objective of this study was to carry out an integrated assessment of agroecosystem health and sustainability with special focus on smallholder farms in the central highlands of Kenya (Chapter 1). The agroecosystem health framework was successfully adapted for use in a smallholder-dominated agroecosystem (Chapter 2). Participatory methods (Chapter 3), systems analyses (Chapter 4), soft systems methods (Chapter 5), and conventional research approaches were combined in an open-ended, adaptive research-and-development process. Two suites of health and sustainability indicators were developed. The first suite, which was community driven, enabled farmers and communities to assess the health and sustainability of their own agroecosystem. The second suite was research based and complemented the community-driven suite. This was used to assess the potential impact of community goals on health and sustainability of the Kiambu agroecosystem. Pulse process models (Chapter 4) were used in these assessments. Correspondence analysis was used to refine the research-based suite of indicators (Chapter 6) as well as to analyze data obtained using indicators (Chapter 7).

8.2 SUSTAINABILITY

Communities' cognitive maps (Chapter 4) and descriptions of their vision of a healthy and sustainable future (Chapter 6) seem to indicate that they perceive sustainability as resulting from accelerated economic development. They do not perceive resource stocks as consumable piles but rather as consisting of renewable and nonrenewable portions, with capacity for regeneration if the system is properly utilized. Their descriptions seem to indicate that they perceive it possible to gainfully and sustainably increase the utilization of their agroecosystems, presumably through the use of technology to realign the way resources are utilized and to support the agroecosystem status and function. The growth concept of sustainability seems to be the most congruent with community perceptions.

The growth concept of sustainability emphasizes a balance among people, their habitat, and economic systems. It assumes that there exists an optimal level of productivity for the agroecosystem, and that successful management involves attaining and sustaining this optimum. This optimum depends on the rate at which resources are regenerated, the rate at which the environment is able to absorb wastes and by-products, and the existence of appropriate technology to facilitate both the exchange of nonrenewable resources for the renewable ones and to support the integrity of the agroecosystem. That communities perceive this to be the case is illustrated by their recognition of the need to enhance both productivity and soil quality through use of manure. Another example is where they attempt to optimize the use of agrochemicals to increase productivity but minimize their perceived negative health impacts on the community.

Communities showed great concern for sustainability issues and had a clear-cut idea of what it meant (Chapter 6) in their own agroecosystem despite its vague and ambiguous definition. This underscores the global appeal of the sustainability concept and its power in stimulating debate on natural resource husbandry. Like the concept of health, sustainability—it seems—is capable of being operationalized without the need for further refinement of its definition. It is probable that refinement, which implies dilution of its holistic connotation, may result in the loss of its global appeal and therefore its potential to evoke and guide the need for change in natural resource management.

Although communities have a strong sense of what is good and what is bad in terms of the health and sustainability of their agroecosystems (evidenced in their problem analysis; Chapter 3), they did not seem to appreciate the need for—or lacked a capacity for—debating, negotiating, planning, and implementing remedial actions (Chapter 5). Based on their approach in selecting indicators (Chapter 6), sustainable development, to them, implied stating long-term goals for the agroecosystem and then building and evaluating short-term and long-term goals based on these. From their perspective, a sustainability assessment involves an evaluation of the probabilities that the desired long-term goals will be attained given the current management practices and agroecosystem conditions. In system terms, this implies that the agroecosystem together with its socioeconomic subsystems must form a holon with integrity, that is, the emergent property of a holon to regulate and organize its own internal structure and function and to mitigate stresses imposed from the outside so that it can perpetuate itself over all foreseeable external fluctuations. A key requirement for integrity is the existence of monitoring and control subunits within the holon, which in turn implies the existence of at least one measure of performance, a criterion of what constitutes good or bad performance, and the remedial action to be taken for each of the possible outcomes. The inability of communities to pursue collective goals, when contrasted with the communities' demand for action subsequent to the initial village workshops, gives validity to this analysis. The request by communities to form village agroecosystem health committees can be interpreted as an attempt to build monitoring and control structures.

8.3 AGROECOSYSTEM HEALTH

While the concept of sustainability evokes notions of natural resource husbandry, the agroecosystem health paradigm provides a compelling framework for the successful management of agricultural and ecological systems. Community members, extension agents, and policymakers in this project used concepts derived from the health disciplines to assess and set goals for their agroecosystem; to debate, negotiate, and plan remedial measures; and then to monitor and evaluate progress. In this regard, sustainability was seen as analogous to health in the sense that they are both objectively definable states of dynamic systems that, once described, can be effectively pursued.

Based on the type of indicators selected by communities, it seems that communities perceived agroecosystem health from a fitness assessment rather than a diagnostic perspective. In the latter perspective, the objective of the process is to discover and characterize pathological processes and the risk factors associated with them. The former focuses on the capabilities of the system and what enhances it. Based on this perspective, the key health attributes are productivity, vigor, resilience, equitability, stability, and integrity. The objective of an agroecosystem health assessment is to understand how the system can achieve and sustain desired community outcomes. In contrast, the objectives based on a diagnostic perspective would be to discover potential risk factors to the attainment of community goals. Important attributes in this case would include equitability, elasticity, inertia, and vulnerability. While the fitness assessment perspective was used by the communities to help them set reasonable goals for their system, ignoring the diagnostic perspective resulted in failure of community action plans after a significant amount of resources had been expended, resulting in a lot of frustration and decline in the communities' capacity for collective action.

Communities were able to develop a reasonably parsimonious suite of indicators. This is at odds with the assertion that ecosystems present an almost-infinite list of potential indicators. This assertion stems from models of agroecosystems as dynamic states of a hard system. In contrast, communities and researchers in this project modeled agroecosystems as problem-based soft systems. Indeed, questions of sustainability and health would have little relevance in systems that do not include some components of human influence. The question of sustainability implies a human activity system and an existence of a complex problem situation. Building problem-based models of agroecosystems limits the choice of indicators to those related to the subsystems in which the problems occur and are manifest. Building problem-based models requires experiential knowledge of the system, emphasizing the importance of community knowledge of their agroecosystem.

8.4 KIAMBU AGROECOSYSTEM

Scarcity of farmland is an important determinant of the nature of smallholder agriculture in Kiambu. This is evidenced by the differences in agricultural practices and productivity among the six villages (Chapter 2). The availability of markets and demand for produce are other important issues, as evidenced by the abandonment of recommended farm enterprises (based on agroecological suitability), such as coffee and tea production in Mahindi and Kiawamgira and sheep production in Thiririka, for those that are largely market driven, such as dairy and vegetable production. While these trends indicate adaptability in a general sense, they could also be reflective of some kind of instability in the system given the relatively short time span over which they are occurring. More important, these communities seem to have a high degree of adaptability regarding the kind of farm enterprises in which they are willing to engage, and their farming decisions appear to be linked to market availability, indicating a high degree of effectiveness.

Although, on average, income for a household was low, many of the households had diverse sources of income, including off-farm employment. It was difficult to assess the relative stability of these incomes, but it can be assumed that the diversification observed is an attempt to minimize risk. On the other hand, there was an oversupply of labor in most of the households, and the diversification may be simply a consequence of this. Interestingly, communities perceived labor as one of the products they export. This was not seen as competing with demands for agricultural production except in Gikabu village, where demand for casual labor in the neighboring tea estates was seen to be in direct conflict with the needs for smallholder tea production. Another interesting aspect was that although communities saw a direct relationship between education/knowledge/skills and access to off-farm employment, they did not appreciate the value of off-farm employment as resulting in increasing community contact with the outside world and as a source of knowledge and information. It is likely that this is the mechanism through which the villages obtain the critical information that has facilitated their adaptation to changing circumstances.

8.5 HEALTH AND SUSTAINABILITY ASSESSMENT

In general, the agroecosystem approach has many attractions from both the research and development perspectives. The health paradigm used is easily understood and conceptually facilitates the diagnosis, treatment, follow-up monitoring, and evaluation of agroecosystems. Because health assessments are value laden, their establishment requires community participation if they are to achieve meaningful and lasting results. In addition, analyses at different holarchical scales are helpful for communities since development requires cooperation across households and villages and larger levels of organization, such as government and other agencies. A key feature of the process is that community organization—manifest as a capacity for collective action—is both a prerequisite and an outcome of the process. While communities with less capacity for action will realize minimal impacts in the short term, the long-term effects will be increased organization—setting the stage for better outcomes in the future.

There are a number of practical implications noted during the project. The first was that this research paradigm allows for the development of an effective forum for community and research collaboration. The second was that integrating participatory and standard research approaches to address community concerns can achieve tangible results. The research input helped communities to better understand the choices to be made in developing and modifying community action plans. For researchers, there were real benefits from communities generating research questions based on the real needs of the community. Research results, in this context, are more likely to be adopted and sustained. Furthermore, the various processes and steps of the framework increase community awareness, self-knowledge, and analytical skills. This, together with the enhanced capacity for action, increases their ability to adapt and hence improve their health.

The main difficulties in the agroecosystem approach are related to its time horizon and location specificity. As the process is open ended, only its initiation and early development fit into a standard project time frame. Longer-term issues, such as clear-cut assessments of sustainability, require longer-term monitoring mechanisms. It is difficult to judge how lessons learned in one set of communities can be generalized to other communities and agroecosystems. In our view, the process is transferable with fairly moderate adaptive changes. Some of the lessons may be instructive in the management and assessment of similar agroecosystems, but this will become clear as more studies of this nature are undertaken, compared, and contrasted.

The holistic approach adopted in this process, while essential to establish the crucial context for decisions and priorities, means that researchers and development agents with narrowly defined terms of reference will not be able or willing to use this approach, especially if their priorities are not strongly linked with those of the target communities. Agroecosystem health assessments can be initiated under the umbrella of agencies (such as the National Agricultural Research Service [NARS]) that have broader terms of reference, while those with narrower focus are best integrated in a secondary process based on the outcomes of the initial assessments and analyses.

The key lesson for communities is that the health approach to community description, problem analysis, and action planning only works if the community is committed to and leads the process. All communities had some success with this approach, mainly related to their organizational ability and commitment. The participatory techniques for analyzing, planning, and monitoring action plans were effective and contributed to community mobilization and action. Communities also discovered that they could learn effectively from the experiences of other communities. Thus, strategies to foster intervillage collaboration need to be an important feature of such efforts. Researchers from all disciplines involved in this project appreciated the ability of communities to formulate "research" questions and analyze constraints. This approach provides an important pathway for developing relevant research questions. In addition, the perspective and ability of communities to analyze their problems were impressive and can be an important tool for researchers trying to assess complex issues using soft systems and more traditional multivariate approaches.

8.6 SUMMARY

The increasing realization that human activities have complex impacts on the health and sustainability of agricultural and ecological systems has led to the increasing interest in holistic and adaptive approaches in the management of human activity systems. The results presented in this study strongly illustrate that such holistic approaches are feasible and demonstrate the potential of the agroecosystem health paradigm as a framework for incorporating these concerns into the decision-making processes of agricultural communities in a tropical highlands agroecosystem.

9 General Conclusions

1. This work demonstrates that a holistic approach to investigating agroecosystem health (AESH) and beginning to implement sustainable processes for AESH improvement is feasible even with complex field situations.
2. Communities were able to use the concept of health to discuss and model approaches to improve their livelihoods. The approach provides a simple, yet highly specialized language—understood by the communities, researchers, extension agents, development agents, and policymakers—for discussing issues relating to AESH and sustainability.
3. Although remarkably similar to traditional methods of integrated community development, the AESH framework is based on the principles of systems theory and practice, participatory and action research methods, and conventional research methods combined into a transdisciplinary framework. The AESH framework as applied in this study is a metaphor to structure how people think about their actions—social or economic—and their implication on the biophysical world to improve their own well-being and to conserve the natural resource base on which their survival depends.
4. A unique feature in this process was that communities, researchers, and development agents played complementary roles. While the communities' role was crucial to understanding the system and in defining the criteria for health, the role of the researchers as experts in methods and that of the development and extension agent as subject experts was critical to the overall success of the project.
5. Cognitive maps, graph theory, and pulse process models were useful in analyzing community perceptions of factors that influence AESH and sustainability. That communities easily understood and applied cognitive maps to depict their perceptions combined with the fact that the cognitive maps were largely in agreement with findings from the participatory workshops indicate the potential of this method. In an action research process, cognitive maps can be reevaluated and updated in each action research cycle as the local theory develops. In this way, they can serve as a record of the developing local theory, an analytical tool, and a means for assessing new goals and objectives.
6. Rich pictures were useful in helping communities to analyze the different perspectives that existed on a problem situation and in guiding consensus building, negotiation, and compromise.
7. Communities used the information derived using indicators to reevaluate their action plans, objectives, and goals. This demonstrates the success of the AESH approach in operationalizing the concepts of AESH and sustainability and in incorporating them into the decision-making processes of the communities.

8. Technical improvements, such as geographical information systems and remote sensing, self-reporting systems, and automated and integrated computer-based data-gathering techniques, would be needed to make indicator data more reliable and cost-effective in the long term.
9. Both community-driven and researcher-based indicators were useful because the two suites provided complementary but fundamentally different information. Because communities often lack the capacity to develop and measure quantitative indicators, while researchers and policymakers lack the knowledge and mandate to make value-based judgments, it seems that decision support systems for such integrated and adaptive approaches as sustainability and AESH should include both components to provide a balanced assessment.
10. MCA (multiple correspondence analysis) was useful in summarizing and presenting indicator data for two reasons: (1) the incorporation of targets and thresholds in the process of categorizing the indicators, thus providing an intuitive interpretation, and (2) projection of data from the initial and subsequent measurements into a multidimensional space.

Appendices

APPENDIX 1: QUESTIONNAIRE USED TO CARRY OUT A CENSUS OF ALL THE LAND-USE UNITS IN THE VILLAGE

1. Village: ______________________________ 2. Date:__/__/98

3. Name of enumerator: ______________________________

Details of informant

4. Name of person answering questions: ______________________________

5. Relationship to the owner of farm:
0. owner 1. friend 2. son/daughter 3. husband 4. wife 5. employee 6. Other. Specify _________

6. Age: 0. Adult (more than 22 years old) 1. Youth (between 15 and 22 years old) 2. child (less than 15 years)

Details of the farm-owner and household

7. Full name of owner: ______________________________

8. Owner is: 0. Male 1. Female

9. Full address: ______________________________

10. Occupation of owner: 0. farmer 1. bussiness man 2. employee

11. Age of owner: 0. Between 30 and 60 years 1. Less than 30 years 2. More than sixty

12. Number of people in the household:
0. Children (less than 15 yrs) ______ 1. Youth (16 to 24) ______
2. Adult (25 to 60)______ 3. Aged (more than 60)______

13. Of these, how many go away to: 0. school: ______________ 1. Work off-farm:_________ 2. Businesses: __________

14. List the names of co-operatives, groups or committees that the owner is a member of:
1. Name______________________ Purpose______________________ use a blank sheet
2. Name______________________ Purpose______________________ If more space is
3. Name______________________ Purpose______________________ needed

Details of the farm-manager

15. Name of person who minds the day-to-day activities of the farm (manager) ______________________________

16. Manager is: 0. Male 1. Female

17. Age of manager: 0. Between 30 and 60 years 1. Less than 30 years 2. More than sixty

18. What is his/her relationship to the owner of the farm?
0. owner 1. son/daughter 2. husband 3. wife 4. employee 5. Other. Specify _________

19. How many years has the manager worked on the farm? ______________________________

20. What is the level of education of the manager? 0. None 1. Primary 2. Secondary 3. College 4. Others. Specify _____

21. Number of people employed (permanently) to work on the farm: ______________________

22. Indicate activities for which casual labourers are employed:
0. None 1. Land preparation 2. Planting 3. Weeding 4. Harvesting 5. Others. Specify ___________

Farm-plan

23. Where do you get water for domestic use? 0. Tap 1. River 2. Well 3. Bore-hole 4. Rain water 5. Others. Specify______

24. What crops are planted in the farm (current season)?
1. Maize/beans/potatoes________ acres 2. coffee ________acres 3. Tea ________acres
4. Napier________acres
5. Kales/cabbages ________acres 6. Others. Specify___________ ________acres

25. What is the total size of the farm? ______________ acres.

26a. List the number of cattle on the farm and their breed:
0. None 1. Zebu :___ 2. Fresian: ___ 3. Jersey: ___ 4. Arshyre: ___ 5. Guernsey: ___ 6. Crosses:___
7. Unknown: ___

26b. Where do you take your milk for marketing? 0. No cow is being milked 1. Co-operative. Name ___________
2. Middlemen. 3. Sell to other people.

27. Indicate the number of sheep on the farm: Male ______ Female________

28. Indicate number of goats on the farm: Male ______ Female________

29a. Indicate type and number of Chicken on the farm: 1. Layers: ___________ 2. Broilers: __________ 3. Others:______

29b. Where do you sell eggs and meat? 0. No chicken are laying 1. Order
2. Locally to other people 3. Others. Specify__________

Farm productivity

30a. How would you rate the productivity of your farm? 3. Very good 2. Satisfactory 1. poor 0. Very poor

30b. If productivity is poor, what is the most important cause?
0. Unknown 1. small farm size 2. low soil fertility 3. lack of skills
4. labour constraints 5. others. Specify _____

31. What is your income from the following farm enterprises:
1. Cash crops (tea, coffee, pyrethrum) ___________ per ___________
2. Food crops (maize beans, potatoes) ___________per ___________
3. Livestock (milk, eggs, sale of stock) ___________ per ___________

32. Which of your farming activity would you like to improve if resources were available?
0. None 1. Cash-crops 2. dairy 3. food-crops 4. Horticulture
5. New enterprise. Specify ___________ 6. Others. Specify _____

26b. If none, why ?____

33. Briefly explain how you would make the improvement ______________________________________

34. What resources would you require to make these improvements? ____________________________

35. What are the likely sources of these resources? ____

APPENDIX 2: QUESTIONNAIRE USED FOR INDICATOR MEASUREMENT AT THE LAND-USE UNIT LEVEL

THE AGROECOSYSTEM HEALTH PROJECT
UNIVERSITY OF NAIROBI

SUBHOLON LEVEL ASSESSMENTS

HOUSEHOLD DETAILS

To be completed for each household

The respondent is the household head (mwene mucii) or the spouse

---

SECTION 1
(Fill-in this section immediately before the interview)

Interviewer's Code: __________ Date: __/__/__

Full name of HHH: ________________________ Code: ________

Village: ________________________________

SECTION 2
(Fill-in this section immediately asking to talk to the HHH (Mwene muci↔))
(circle the answer that corresponds to response given)

1. *HHH is:*
   i. *Male*
   ii. *Female*

2. *If respondent is not HHH, indicate name of respondent*________________________

3. *Available respondent is*
   i. *household head*
   ii. *Spouse of Household head*
   iii. *Other. Specify*____

4. *If respondent is not HHH, give reasons why* ______________________________
____
____

SECTION 3
(Inform the respondent that you will ask some lengthy questions and that they can stop you at any time if they feel that they need rest. Select a comfortable place for the interview)

5. For the household head, his or her spouse and their offspring, give details of their ages, occupation, education level and marital status

Code	Names	Date of birth	Sex	Occupation	Monthly Farm Income[1]	Monthly Non-farm Income1	Level of Formal education	Other training/ courses	Can the individual read a newspaper	Religion	Does the individual eat and live in this household? Y/N	Marital status	Date of marriage
101													
102													
103													
104													
105													
106													
107													
108													
109													
110													
111													
112													

1 Non-firm income is income from non-agricultural activities

6. List the names of other people not mentioned above who live and eat in this household and give details of their ages, occupation, education level and marital status.

code	Full names of person	Date of birth	Sex	Occupation	Farm Income	Non-Farm Income	Formal education level	Other training /courses	Can read a newspaper? (y/n)	Religion	Relation with household head	Marital status	Age at marriage
201													
202													
203													
204													
205													
206													
207													
208													
209													
210													

7. If anyone in this household has been ill or injured within the last 30 days from today, please indicate the type of injury or illness, how serious it was and what action was taken.

Code number	Description of illness or injury	How many days did illness last	How serious (1=very, 2=moderate, 3=not very)	Action taken 1= none 2= hospital 3=shop medicine 4 = other (specify)

8. If anyone in this household has been admitted to hospital in the last calendar year, please indicate the reasons for admission, the duration of hospitalisation and the outcome

Code number	Reason for admission 1= Severe illness 2= Surgical operation 3=accident/injuries 4= other (specify)	How many days was the person hospitalised	Name of Hospital	Outcome 1= recovered 2= still in hospital 3=home but ill 4 = other (specify)

9. Give an estimate of the amount of money you spent in the last calendar year for medical services. ______________Ksh

10. If any of your family members has a permanent disability (□onje wa m□themba), please give details ______________________________

11. If there is a household member(s) who has died in the last calendar month, please indicate the cause(s) of death ______________________________

12. If there is a household member(s) who has died in the last calendar month, please indicate the age at time of death ______________________________

13. For each child below 5 years of age in this household, please allow us to examine the Child Health Card

Code Number	Does Child Health card exist?2	Name of health facility	date of birth	Number of vaccination events	Number of weight measurements	Last weight measurement (Kg)	Age (months) at last weighing

Comments

2 If Child Health Card does not exist, only the date of birth is required.

14. For the following types of foods indicate **(tick)** those which are used as food in this household within the last calendar year. Also indicate which ones were available in your farm.

Type of food	Used as food	Purchased	Available in farm	Type of food	Used as food	Purchased	available in farm
Maize				Carrots			
Beans/peas				arrow roots			
Kales				Fruits			
Eggs				Sweet potatoes			
Fish				Irish potatoes			
Milk				Cassava			
Pumpkins				Millet sorghum			

15. Would you say there have been significant changes in diet in your lifetime?
   i. Yes ______
   ii. No ______

16. If there has been a significant change of diet in your lifetime, has the change been for the better or worse?
   i. For the better. Explain ______
   ii. For worse. Explain ______
   iii. Neither good nor bad

17. If there has been a significant change in diet, what kind of change has occurred?
   i. Frequency of meals in a day
   ii. Food types
   iii. Methods of preparing the food
   iv. Others. Specify ______

18. If there has been a significant change in diet, what would you say are the most important causes of these changes?

   i. Changes in our ability to buy food (change in prices or incomes)
   ii. Changes in the types of food produced in our farms
   iii. Commodities no longer available
   iv. Others. Specify ______

19. List all the buldings in this homestead, indicating the area they cover , what they are made of and their use.

Name of buildings	Use of building	Area covered by the building	Construction materials			Distance from main House[3]
			Roof	Walls	Floor	

[2]**The structure in which the household head sleeps**

20. For each of the following resources indicate their availability in the homestead, and indicate who should be consulted before they are sold, used or lent out.

Resource	**Person consulted before USE** 1= not available 2=no consultation 3= Husband/father 4=wife/mother 5= others (specify)	**person consulted before LENDING** 1= not available 2=no consultation 3= Husband/father 4=wife/mother 5= others (specify)	**Person consulted before SALE** 1= not available 2=no consultation 3= Husband/father 4=wife/mother 5= others (specify)
1. Land/plots			
2. Car			
3. Livestock			
4. Cash crops			
5. Food crops			
6. Household goods			
7. Children *use=labour* *lending = sending* *sale = allowing him to stay with someone else*			
8. Bicycle			
9. Non-farm income			
10. Bank account			

21. What is the distance from your house to the nearest road that has transport _______ Km

22. For the following amenities indicate the location and the distance from the homestead

Amenities	Name	Within Walking Distance? Y/N	Distance/Time taken
1. Dispensary			
2. Hospital			
3. Primary school			
4. Police			
5. Shopping center			
6. Market			
7. Chief=s office			

23. How much did you spend on water in the last calendar month? __________ Ksh.

24. What is the cost per litre? ________________________ Ksh

25. What is the distance to the nearest river/natural spring _________Kms.

26. Did you use water from a natural spring or a river for domestic purposes in the last calendar year?
    i. Yes
    ii. No **(Go to question 29)**

27. If you used water from a natural spring or a river for domestic purposes, indicate month(s) when used ____________________________

28. In the last calendar year, did you use water from a river or a natural spring for agricultural purposes?
    i. No
    ii. Yes. for livestock
    iii. Yes. for crops

29. Do you have rainwater-harvesting facilities?
    i. Yes
    ii. No **(go to question 31)**

30. If you have rainwater-harvesting facilities, how long does the harvested water last after the rainy season? _______ weeks.

31. If you have access to piped water, indicate how often the taps have water
    i. no piped water
    ii. once a week
    iii. some days in the week. Specify __________________________
    iv. some day in the month. Specify __________________________
    v. Others. Specify _______________________________________

32. If you have access to piped water, how much did you pay last month for the water? ____________________ Ksh

33. Do you use water from a shallow well (irima/ g↔thima) for domestic purposes?
    i. Yes
    ii. No

34. <u>If you have a shallow well</u> in your homestead, what materials have you used to cover the well?
    i. No well
    ii. no cover
    iii. wooden cover
    iv. Metal cover
    v. concrete cover
    vi. other Specify ______________________________

35. <u>If you have a storage tank</u> for water, indicate its storage capacity ________litres

36. How many people share water from the tank? ______________

37. <u>If you have a storage tank</u>, what is it made of?
    i. iron sheets
    ii. concrete
    iii. plastic or fibre glass
    iv. others. Specify

38. Do you have different uses for water from the different sources
    i. Yes. Explain ______________________________
    ii. No

39. List the different kinds of fuels used in your household in the last calendar year. Also indicate the main source, frequency of use and the cost per month

Type of fuel	Use 1=lighting 2=cooking 3=others.(specify)	Frequency of use	Cost per month (Ksh)

40. If firewood was used as fuel in your household in the last calendar year, where did you obtain it from?
    i. Woodlot on my farm
    ii. Prunnings from tea and/or coffee
    iii. Purchased
    iv. Gathering outside the farm
    v. Other. Specify ______________

41. What alternative sources of fuel for cooking and lighting (energy technologies) do you know of (eg. biogas, Kuni Mbili, solar cooker, fireless cooker, battery, solar panel)?

____

____

42. List the groups, associations, organizations, clubs or societies of which you are a member. For each, indicate the objective or purpose.

Name of group	**Purpose or objectives of group**

43. For each member of your household, give a general overview of the usage of time yesterday, Saturday and Sunday. *(All the details can be provided by the respondent unless clarification is needed. Individuals with similar routines to be entered in the same row but the individual codes given. Make sure that everybody is accounted for in all the tables).*

1. Day(s) of the week ______________

Codes	Early morning	Mid morning	After noon	Evening

2. Day(s) of the week ______________

Codes	Early morning	Mid morning	After noon	Evening

3. Day(s) of the week ______________

Codes	Early morning	Mid morning	After noon	Evening

4. Day(s) of the week ______________

Codes	Early morning	Mid morning	After noon	Evening

44. Apart from what you get from the farm and the support from members of this household, is there another regular source of support, income or resources for this household? Please specify form of support

		Form of support
i.	None at all	
ii.	From husband working away from home.	______
iii.	From son(s) living away from home	______
iv.	From daughter(s) living away from home	______
v.	Bursaries	______
vi.	Donors	______
vii.	Friend(s)	______
viii.	Other. Specify ______	______

45. How often do you visit friends/Relative who live outside the district?

Friends	Relatives
i. Very often	i. Very often
ii. Often	ii. Often
iii. Rarely	iii. Rarely
iv. Not at all	iv. not at all
v. No friends to visit	v. No relatives to visit

46. How many times did you participate (receive or give) an Itega outside this village in the last calendar year?
  i. Once
  ii. Twice
  iii. Meetings are issue driven
  iv. Other, Specify ______

47. How often were clan meetings (mʃcemanio wa nyʃmba) held in the last calendar year?
  i. We do not hold clan meetings
  ii. I do not belong to a clan
  iii. Once a year
  iv. Twice a year
  v. Once a month
  vi. When need arises
  vii. Other, Specify ______

48. In your opinion, when should parents stop supporting their offspring materially? (k□□r□gam□r□ra/kw□ikaria)
  i. never
  ii. after circumcision
  iii. it depends on sex of individual. Explain ______
  iv. after finishing / dropping out of school
  v. after he or she gets married
  vi. after he or she gets a job
  vii. At a specific age. Explain ______
  viii. Others. Specify and explain ______

Briefly, explain how you would go about discontinuing material support for your daughter

______

49. Briefly, explain how you would go about discontinuing material support for your son

______

50. When do you (parent) stop allocating duties to your offspring?
  i. never
  ii. after circumcision
  iii. it depends on sex of individual. Explain ______
  iv. after finishing / dropping out of school
  v. after he or she gets married
  vi. after he or she gets a job
  vii. At a specific age. Explain ______
  viii. Others. Specify and explain ______

51. Who deals with issues of discipline in the house for the following categories?
  i. Children under 12 years of age ______
  ii. Girls 12 years of age or older ______
  iii. Boys 12 years of age or older ______

52. What form does this discipline take?
  i. Canning / beating
  ii. Refusal of food
  iii. It depends on the person being disciplined. Explain ______
  iv. Other. Specify ______

53. How much money did you spend last year on education (fees, transport, books) for children in the following categories

i.	Nursery school	Ksh. ______
ii.	Primary school	Ksh. ______
iii.	Secondary school	Ksh. ______
iv.	College / University	Ksh. ______
v.	Others. Specify ______	Ksh. ______

54. List the children below 19 years of age in this household who are not attending school. Indicate reasons, last class attended and the age when he/she stopped school.

code	last class attended	age when (s)he stopped school	reasons for stopping

55. List, in the order of priority, the occupations you would have preferred most for your children

Male children	
1	
2	
3	
4	

Female children	
1	
2	
3	
4	

Please indicate the degree to which you agree or disagree with the following four statements

56. I do not see much difference in the lifestyles of those who are educated and those who are not
    i. Agree very much
    ii. Agree
    iii. Disagree
    iv. Disagree very much

57. All one needs is to be able to read and write
    i. Agree very much
    ii. Agree
    iii. Disagree
    iv. Disagree very much

58. Education makes one able to cope better with the challenges of life
    i. Agree very much
    ii. Agree
    iii. Disagree
    iv. Disagree very much

59. It is better to educate a male child than a female child
    i. Agree very much
    ii. Agree
    iii. Disagree
    iv. Disagree very much

SECTION 4

(In the space provided below make any comments on questions asked and any observations that you think affected the responses)

FARM DETAILS

To be completed for each farm
The respondent should be the person who decides on the day-to-day operations of the farm

Name of respondent:______________________________
Name/Code of household head______________________________

Interviewers' Code______________________________

1. Respondent is :
   i. Manager
   ii. Owner/household head
   iii. Spouse
   iv. Other / specify: ______________________________

2. Give reason(s) why manager is not the respondent ______________________________

3. What is the total area of land that you use for agricultural production?____________acres

4. How much of this land belongs to you?[1] ______________acres

5. How much of this is <u>rented from somebody else</u>? ______________acres

6. How much of your land have you <u>leased out</u>? ______________ acres

7. Name of manager ______________________________

8. Manager's relation to household head

   i. owner
   ii. son/daughter
   iii. husband
   iv. wife
   v. employee
   vi. Other. Specify ____________________

9. Sex of manager:

   i. Male
   ii. Female

10. What is the age of the manager : _________years

11. For how many years has the manager been involved in farming activities? ________ years

12. Did the manager get advice on farming from sources outside the household during the last calendar year? (circle appropriate responses):

    i. No. (go to question 16)
    ii. Yes. Relative/neighbor
    iii. Yes. Other farmers
    iv. Yes. Agricultural extension agents
    v. Yes. Livestock extension agents
    vi. Yes. Animal health assistant / Veterinarian (daktari)
    vii. Yes. Company sales rep (crops)
    viii. Yes. Company sales rep (livestock)
    ix. Yes. Dairy Co-op rep
    x. Yes. Coffee or Tea co-op rep
    xi. Yes. Farmers training centres
    xii. Yes. Research stations
    xiii. Yes. Other (specify) ____________________________

13. Are the extension messages easy to understand?

    i. Yes
    ii. No

14. <u>If you received information from outside</u>, how much money did you spend to obtain it in the last calendar year?_____Kshs.

15. How useful was the information obtained?

    i. Not useful
    ii. Very useful
    iii. Useful

16. Do you employ non-family members on your farm? If so, how many?

    i. No non-family employees
    ii. Yes. Number of non-family employees ____________________

17. For what periods of the year do you employ non-family workers?

    i. all year round
    ii. harvest time
    iii. according to the need/work load
    iv. other. Specify:____________________________

[1] **ownership of land means that the person has a title deed.**

18. List the crops grown in this farm over the last calendar year (Jan to Dec 1998). Indicate the acreage, quantity planted and quantity harvested during each cropping season

Code	Crop	Month of harvest[2]		Acreage (acres)	Quantity planted[3]	Harvest (Kg)[4]	Quantity sold (Kg)
01		1					
		2					
		3					
02		1					
		2					
		3					
03		1					
		2					
		3					
04		1					
		2					
		3					
05		1					
		2					
		3					
06		1					
		2					
		3					
07		1					
		2					
		3					
08		1					
		2					
		3					
09		1					
		2					
		3					
		2					
		3					

19. For each of the following pair of crops indicate which was the most profitable (in terms of income generated) in the last calendar year?

code/name	01	02	03	04	05	06	07	08	09
10									
09									
08									
07									
06									
05									
04									
03									
02									

20. For each of the following pair of crops indicate which was the most costly to produce in terms of time, money and labour?

code/name	01	02	03	04	05	06	07	08	09
10									
09									
08									
07									
06									
05									
04									
03									
02									

21. List the crops that are currently being grown on your farm

Crop (name)	Acreage (acres)	Quantity planted (kg or # of plants)	Expected yields (Kg)

[2] If harvest is all year round, indicate annual totals
[3] Kg of seeds. For coffee, tea indicate number of plants harvested from
[4] Includes all produce harvested for sale and for home consumption

22. For each of the crops, indicate the most common pest and disease problems[5] that occurred in the last calendar year, their causes and the action taken

crop / code	Pest/Disease/symptom[5]		Cause	Action taken
01	1			
	2			
	3			
02	1			
	2			
	3			
03	1			
	2			
	3			
04	1			
	2			
	3			
05	1			
	2			
	3			
06	1			
	2			
	3			
07	1			
	2			
	3			
08	1			
	2			
	3			
09	1			
	2			
	3			
10	1			
	2			
	3			

23. Indicate the types and quantities of fertiliser and pesticides that you purchased in the last calendar year (Jan 1998 – Dec 1998)

Type F = fertiliser P = pesticide	Name	Used for which crops?	Total quantity purchased (Kg or litres)	Total cost (Ksh)	Total amount used (kg or litres)

24. List the number of cattle on the farm that are being milked ________________

25. What is the total amount of milk that was produced by cows on this farm yesterday? ____ kg

26. Of the milk produced yesterday, how much was sold? _____ kg

27. Where was the milk sold?
   i. No cow is being milked
   ii. Co-operative. Name ____________________________
   iii. Middlemen.
   iv. To neighbours in the village.
   v. Individuals or businesses outside the village

[5] Use the local name and/or the farmers description to identify the disease

28. For each of the cows being milked, indicate the date (month and year) when it gave birth and the amount of milk it produced yesterday

Cows ID (name or description)	Date of last calving	Milk produced yesterday (kg)

29. For each of the following categories of livestock indicate the number currently on the farm, and the number that has been newly introduced

	Category	Goats (Mbuzi)	Sheep (Kondoo)	Cattle	Pigs
1	Adults Females currently on the farm				
2	Adult males currently on the farm				
3	Number of live births in the last calendar year				
4	Number of stillbirths and abortions in the last calendar year				
5	Total number of animals sold or given out in last calendar year				
6	Total number of animals dead (exclude stillbirths and abortions)				
7	Total number of animals brought into farm (purchases, gifts or exchanges)				

30. For each of the following animals indicate the diseases[6] that were the most common in the last calendar year, their cause, and the action taken

Type of Livestock	**Diseases/symptoms[6]**		**Causes**	**Action taken**
Cattle	1			
	2			
	3			
Pigs	1			
	2			
	3			
Sheep	1			
	2			
	3			
Chicken	1			
	2			
	3			
Goats	1			
	2			
	3			

31. How do your animals get access to forage (malisho)?
   i. grazing/pasture
   ii. cut or purchased and transported to animals (zero grazing)
   iii. combination of the above.
   Explain:________________________

32. <u>If your animals graze or are on pasture</u>, how far from the night housing/shelter do they travel?
   i. <0.5 km
   ii. 0.5 - 1 km
   iii. >1 - 2 km
   iv. >2 - 4 km
   v. >4 - 6 km
   vi. >6 km

33. <u>If your animals ARE grazing or are on pasture</u>, how much area do they have access to? ___________(acres)

34. <u>If your animals graze or are on pasture</u>, are they ever close to areas where "wild animals" are seen? If so, which wild animals?

[6] Record the descriptions as given by the farmer

i. no wild animals in area
ii. buffalo (mbogo)
iii. bush buck/"antelopes" (Thwariga)
iv. other. Specify:______________________

35. Does your housing/grazing routine differ between wet season and dry season?
i. No
ii. Yes. Describe:

36. If your animals DO NOT GRAZE, how much fodder do you provide per day?
______ Kg .

37. If your animals DO NOT GRAZE, compare the sources of fodder in each of the following pairs and indicate which one you obtained the highest amount of fodder from in the last calendar year.

Source	1 Other sources. Specify	2 purchased grass /hay	3 grass from public land	4 Purchased crop residues	5 crop residues from farm	6 Purchased napier
7 Napier grown on-farm						
6 Purchased napier						
5 Crop residues from farm						
4 Purchased crop residues						
3 Grass/hay from public lands						
2 purchased grass/ hay						

38. What do you do with the manure from the housing area/pens/corral?
i. used as fertilizer on crops
ii. sold
iii. used as fuel
iv. stored but not used
v. other. Specify:

39. To which animals do you provide commercial feeds (dairy meal, bran, pollard, maize germ etc)?

Animal	**Amount (kg) per day**
Cows - immediately after giving birth	
Cows that are being milked	
non-milking cows	
adult bulls	
calves (less than 1yr)	
Adult sheep and goats	
kids (less than 1yr old)	
lambs (less than 1 yr old)	
Young pigs (less than 1 yr)	
Adult pigs	
non-laying chicken	
Laying chicken	

40. Which of the following materials do you offer to your livestock?

Material	Cost per unit	unit
Brewers mash (Machicha)		
Pyrethrum husks (beniku)		
Fish meal		
blood or bone meal		
mollasses (Cukari wa nguru)		
Chicken manure		
others. specify		

41. Indicate which animals receive these materials and the amount per day

Animal	Material	Amount (kg) per day
Cows - immediately after giving birth		
Cows that are being milked		
non-milking cows		
adult bulls		
calves (less than 1yr)		
Adult sheep and goats		
kids (less than 1yr old)		
lambs (less than 1 yr old)		
Young pigs (less than 1 yr)		
Adult pigs		
non-laying chicken		
Laying chicken		

42. Do you use any of the following as supplements? If so what type? *(circle all appropriate responses)*

i. no other supplements
ii. salt lick
iii. salt and mineral mix
iv. vitamins
v. antibiotics
vi. other. Specify:

43. If you feed supplements, to which animals do you offer which supplement?

Animal	Supplement
1) Cows – immediately after giving birth	
2) Cows that are being milked	
3) non-milking cows	
4) weaned heifers	
5) weaned bulls	
6) calves < 2 yr of age	
7) kids	
8) lambs	
9) Adult sheep & goats	

44. What is the total number of broilers chicks that you bought in the year?
____

45. What is the total number of broilers that you sold in the year? __________

46. What was the average price per kg of broiler that you sold? __________

47. What is the total number of Layer chicks that you bought in the year? __________

48. What is the total number of Layers that you sold in the year? __________

49. What was the average price per kg of layers that you sold? __________

50. In the last calendar year, how many times did you
a) obtain veterinary services ? ________ times
b) purchase any kind of remedy for your livestock? ________ times

51. If you obtained any veterinary services in the last calendar year, how much did it cost you in total? ________ Ksh

52. If you purchased any kind of remedies for your livestock in the last calendar year, how much money did you spend in total ________ Ksh

53. How would you rate the productivity of your farm?
i. Very poor
ii. poor
iii. Satisfactory
iv. Very good

54. If productivity is poor, what is the most important cause?
i. Unknown
ii. small farm size
iii. low soil fertility
iv. lack of skills
v. labour constraints
vi. lack of capital
vii. poor infrastructure
viii. lack of market
ix. others. Specify ________

55. How much money did you spend last calendar year on the following farm enterprises (interviewer to assist the farmers to make an estimate. Include fertilisers, chemicals, transport and labour)
i. Cash crops (tea, coffee, pyrethrum) ________ per ________
Transport ______ Fertiliser ______
Pesticides ______ Labour ______
Seeds ______ Other. Specify ______ ______
ii. Food crops (maize beans, potatoes etc) ________ per ________
Transport ______ Fertiliser ______
Pesticides ______ Labour ______
Seeds ______ Other. Specify ______ ______
iii. Livestock (milk, eggs, sale of stock) ________ per ________
Transport ______ Feed ______
Pesticides ______ Stock ______
Other. Specify ______ ______

56. How much money did you earn in the last calendar year from the following farm enterprises (interviewer to assist the farmer to make an estimate, include only direct sales)
i. Cash crops (tea, coffee, pyrethrum) ________ per ________
ii. Food crops (maize beans, potatoes etc) ________ per ________
iii. Livestock (milk, eggs, sale of stock) ________ per ________

57. Which farming activities would you like to increase or start if resources were available?
i. None (Go to 62)
ii. coffee
iii. tea
iv. dairy farming
v. sheep and goats rearing
vi. pig rearing
vii. poultry farming
viii. food crops for home consumption

ix. food crops for sale. specify ____________
x. non-food crops for sale. specify ____________
xi. Others . specify ____________

58. If you would like to increase your farming or start a new enterprise, give reasons why you have not done so yet ______________________________

59. If you would like to increase your farming or start a new enterprise, briefly explain how you would make that change:

60. What resources would you require to make these improvements?

61. What are the likely sources of these resources?

62. If you would not want to increase your farming or start a new farming enterprise, give the reason(s) why?

63. Did you receive credit (loan) of any form (cash, materials, services) in the last calendar year? If yes please indicate the form of credit you received
i. None (go to 66)
ii. Materials. Specify
iii. Cash
iv. Services . Specify

64. If yes, please indicate the source of credit *(circle as appropriate)*

i. relative
ii. friend
iii. trader/ businessman
iv. bank
v. local co-operative
vi. merry-go-round (mwethia)
vii. Agricultural Finance Corporation
viii. Other (specify) ____________

65. If you received cash credit, indicate what you used the money for? (circle as appropriate)

i. invested in dairy operation
ii. invested in production of food crops for sale
iii. invested in production of crops for home consuption
iv. invested in production of non-food crops
v. invested in other farming. Specify ____________
vi. used to pay household expenses
vii. Other uses. specify:____________

66. If you did not receive any credit in the last calendar year, give reasons why

67. If you had the resources to start a new farm enterprise, which one of the following would you choose to start?

Code/ Enterprises	1 One that would give me higher profit aft the sale of produce	2 One that would give me a regular income	3 One that would meet the food needs of my household	4 One where the chances of making a loss are low
5 The amount of time, money and labour required are low				
4 One where the chances of making a loss are low				
3 One that would meet the food needs of my household				
2 One that would give me a regular income				

Index

A

Achievements, 183, 187, 193
Action planning, 69
Action research, 58
Adaptability, 10
Agriculture, 42–48
  dependence on nonrenewable sources, 3
  expansion, 3
  land-use and, 2
  natural resources and, 2
  practices, environmental degradation and, 3
  productivity, 3
  subsidies, 3
Agroecosystem(s)
  assessment, 33
  defined, 6
  health, 5–13, 198–199
    approach to, 52–53, 203
    assessment and implementation of, 10–11
    concept of, 8–10
    goals, 26
    indicators of, 11–13
    management, 25
  research strategy and methods, 27–34
    indicators, 32–33
    sampling study sites, 30–31
    secondary data and holarchical scales, 29–30
    systemic description and action planning, 31–32
Amplitude, 9
Application, 177–178
Arithmetic analysis, 93
Assessment, 4–5, 10–11, 33
  community-based, 33
  defined, 33
  objective, 2
  research-based, 33

B

Behavior, 91

C

Cognitive map(s), 70, 93–95, 99
  construction, 122
  defined, 92
  usefulness, 92
Collaboration, 58
Community
  defined, 60
  identity, 60–61
  initial contact with, 61
Community-based indicators, 70
Community participation, 58
Community training, 34
Constancy, 9
Cyclical stability, 9

D

Daily activity charts, 67
Digraph
  signed, 95, 122–123
  weighted, 96

E

Ecosystem, 6
  characteristics, 6, 7
  vital signs, 12
Effectiveness, 8
Efficacy, 8
Efficiency, 8
Elasticity, 9, 10
Emancipatory action research, 59
Emergent properties, 7
Equitability, 9
Ethnobiography, 65
Evaluations, 183, 187, 193

F

Focus group discussion, 67
Forest, 2
Fresh water, 2

G

Geometric analyses, 93, 95–96, 123–124
Gikabu, 37, 113, 116–117, 119–122
  actions planned by, 84, 85–86
  agricultural products, 42
  agroecosystem, factors influencing, 117
  agroecosystem health, 99, 100

community goals, 118–119
achievement of, 192
diseases and health concerns, 80, 81, 82, 88
expected outputs, 194
farm income, 44, 46
geoclimatic features, 38
historical background, 71, 73–74
land-use unit, 38, 39, 48, 50
main products, 47
medical facilities, 44
off-farm employment, 38, 39, 41
sheep and goat, 42
transect walk and semistructural interviews, 77–79
trend lines, 72, 75, 76
water supply, 45
wealth and well-being, 79
Gitangu, 35 40, 104–109, 188
achievement of goals, 192
actions planned by committees, 84, 85–86
agricultural products, 42
agroecosystem, 106
health indicators, 156, 158
constraints to productivity, 48
decision making institutions, 72
demographic features, 39
diseases and health concerns, 80, 82
expected primary and secondary outputs, 194–195
farm income, 44, 46
geoclimatic features, 38
goals of respondents, 50
historical background, 37, 71, 73–74
indicator evaluation, 168, 170
land allocation to fodder, 40
land-use units, 38
main products, 47
medical facilities, 44, 158
off-farm employment, 41
prioritization of problems, 83
relationship between community and administration, 71
resources, 41, 42, 43
risk of diseases, 81
sheep and goats, 40, 42
transect walks and semistructural interviews, 77–79
trend line diagrams, 75
water project, 131–132, 136, 140, 143
wealth, 76, 79
Githimia, 35 40, 37, 104–109, 188
achievement of goals, 192
actions planned by committees, 84, 85–86
agricultural products, 42
agroecosystem, 106
cognitive map of factors influencing, 99
health indicators, 157
constraints to productivity, 48
decision making institutions, 72
demographic features, 39
diseases and health concerns, 80, 82
expected primary and secondary outputs, 194–195
farm income, 44, 46
geoclimatic features, 38
goals of respondents, 50
historical background, 73–74
indicator evaluation, 169
land allocation to fodder, 40
land-use units, 38
main products, 47
medical facilities, 44
off-farm employment, 41
prioritization of problems, 83
relationship between community and administration, 71
resources, 41, 43
risk of diseases, 81
sheep and goats, 41, 42
transect walks and semistructural interviews, 77–79
trend line diagrams, 75
water supply, 45
wealth and well-being, 79
Grasslands, 2

H

Health assessment, 49, 53–54, 68, 200–201
Health promoters, 12
Historical background, 65
Holarchical scales, 36, 49–51
Holarchy, 7
Holon, 91

I

Indicator(s), 11–13, 32–33, 147
community-driven, 32, 148, 149–151, 153–155, 158–159
desired qualities, 148
evaluation, 33
feasibility, 151
measurements, 152–153, 158–167, 174, 176–177
questionnaire for, 208–222
potential, 15–16
refinement, 158–167, 174, 176–177
research-based, 32–33
research-proposed, 32–33, 148, 151–152, 158, 172–173, 175–176
spatial trends in, 182–183, 184–185, 190, 193
temporal trends in, 182–183, 185–187

selection, 12–13, 147
suites, 167, 181
comparison of, 170–171
sustainability, 150
use of, 26
validity, 148, 151
Inertia, 9
Influence diagrams. *See* Cognitive maps
Integrity, 7, 10
Intense study sites, 61
Intervention
community-based development, 34
implementation of, 33–34, 49

J

Justification, 16–17

K

Kiambu District, Kenya, 2
actions planned by communities in, 84, 85–86
agroecosystem, 13–15, 199–200
external environment of, 13–14
goals of, 15
hierarchical structure of, 29
holarchical structure of, 36, 51
indicators of health and sustainability, 155, 156–158, 160–161, 165–166, 169–170, 174, 175, 176
internal environment of, 14–15
stratification of, 31
allocation of land resources to various crops, 43
area, 14
community inertia, 136, 140
concerns, priority of, 83
coping with, 82
decision making institutions, 72
diseases and health concerns, 80, 81, 88
factors increasing risk of disease, 81
health and well-being of households, 79
inadequate extension services, 134–136
key resources, 42
land use, 172
medical facilities, 40, 44
Ministry of Agriculture, 142
Ministry of Public Works, 137
off-farm employment, 41
population density, 14
road conditions, 48
sheep and goats, 42
size and location, 34, 35
study sites, 34, 35, 40
water supply, 45
Kiawamagira, 109–110, 111–112
actions planned by committees, 84, 85–86
agricultural products, 42, 45
agroecosystem, factors influencing, 110, 113, 117
agroecosystem health indicators, 158
constraints to productivity, 48
decision making institutions, 72
demographic features, 38, 39
diseases and health concerns, 80, 82
expected primary and secondary outputs, 194–195
farm income, 44, 46
geoclimatic features, 38
goal achievements, 192
historical background, 73–74
indicator evaluations, 168
land-use unit, 38, 39, 48, 50
main products, 47
medical facilities, 44
off-farm employment, 41
prioritization of problems, 83
resources, 41, 43
risk of diseases, 81
sheep and goats, 42
transect walk and semistructural interviews, 77–79
trend lines, 75
water supply, 45
wealth and well-being, 79

L

Local theory, 59
Loop models. *See* Cognitive maps

M

Mahindi, 37, 110, 112–113, 114–116
actions planned by committees, 84, 85–86
agricultural products, 42, 45
agroecosystem, factors influencing, 113, 117
agroecosystem health indicators, 158
constraints to productivity, 48
decision making institutions, 72
demographic features, 38, 39
diseases and health concerns, 80, 82
expected primary and secondary outputs, 194–195
farm income, 44, 46
geoclimatic features, 38
goal achievements, 192
historical background, 73–74
indicator evaluations, 170
land-use unit, 38, 39, 48, 50
main products, 47

medical facilities, 44
off-farm employment, 38, 39, 41
prioritization of problems, 83
resources, 41, 43
risk of diseases, 81
sheep and goat, 42
transect walk and semistructural interviews, 77–79
trend lines, 75
water supply, 45
wealth and well-being, 79
Malaria, 88
Mapping
institutional, 64–65
participatory, 63–64
cognitive, 94–95
Messy situations, 129
Monitoring, 33
Multiple correspondence analysis, 148, 204

P

PAR. *See* Participatory action research (PAR)
Participatory action research (PAR), 58
defined, 58
Participatory research appraisal (PRA), 58
defined, 58
development of, 58
functions, 59–60
goal, 58
PRA *vs.*, 58
process and methods, 60–69
village workshops, 60
Path, 95
Persistence, 9
PRA. *See* Participatory research appraisal (PRA)
Practicality, 177–178
Preference ranking, 69
Presentation and analysis, 67–68
Problem analysis, 69
Problem identification and ranking, 68–69
Productivity, 9
Pulse process models, 96–98, 124

Q

Questionnaires, 206–222

R

Resilience, 10
Risk factors, 12
Root definition, 128, 129

S

Scientific fertility farming, 153
Seasonal calendars, 66–67
Self-sufficiency, 9
Semistructured interviews, 66
Signed digraph, 95
Smallholder farms, 2
Soft system methods, 127–145
models, 139
process, 129–130
results, 130–139
Spaghetti diagrams. *See* Cognitive maps
Stability, 9
assessment of value, 124–125
Study sites, 37
Sub-Saharan Africa (SSA)
decline in agricultural productivity, 3
food demand, 3
Sustainability, 4–5, 197–198
assessment, 4–5, 49, 200–201
definitions, 4–5
implementation, 4–5
indicators, 150
long-term, predictions on, 182
Sustainable agriculture, 2, 4
current definitions, 2
indicators, 2
objective assessment, 2
System, 91
System theory, 91–126
application in villages, 98–122
Gikabu, 113, 116–117, 118–119
Gitangu, 104–109
Githima, 99–104
Kiawamagira, 109–110, 111–112
Mahindi, 110, 112–113, 114–116
Thiririka, 117, 119–121
Systemic description(s), 38–49, 51–52
agriculture, 42–48
demographic features, 38
geoclimatic features, 38
resource use and distribution, 38–42
Systems thinking, defined, 91

T

Thiririka, 37, 117, 119–122
acreage per land-use unit, 38, 39
acroecosystem health indicators, 157, 159
acroecozones, 40
actions planned by, 84, 85, 86
agricultural products, 42
cognitive maps, 99
constraints to productivity, 48

- diseases and health concerns, 80, 81, 82
- farm income, 44, 46, 51
- flooding and, 49
- geoclimatic features, 38
- historical background, 71
- income from cash crops, 40
- land allocation to fodder, 40
- land size, 38
- main products, 47
- negative impact relationships, 99
- progress in goal achievements, 192, 193
- resources, 43
- sheep production, 199
- transect walk and semistructural interviews, 77, 78, 79
- trend line diagrams, 75
- water project, 142
- water supply, 45
- waterlogging and, 49
- wealth and well-being, 79

Time lines, 65
Trajectory stability, 9
Transect walks, 66
Trend lines, 65–66
- Gikabu, 72, 75, 76
- Gitangu, 75
- Githimia, 75
- Kiawamagira, 75
- Mahindi, 75
- Thiririka, 75

V

Vigor, 10
Village
- initial participatory workshops, 61
  - discussion of results from, 84–89
  - follow-up of, 69–70
  - learning tools for, 62–69
  - mapping, 63, 64
  - results, 70–84
  - techniques of, 63
  - topics covered in interviews during, 64

Vulnerability, 10

W

Waruhiu Farmers Training Institute, 144
Wealth ranking, 68
Weighted digraph, 96
World population, 2

For Product Safety Concerns and Information please contact our EU representative GPSR@taylorandfrancis.com Taylor & Francis Verlag GmbH, Kaufingerstraße 24, 80331 München, Germany

Printed and bound by CPI Group (UK) Ltd, Croydon, CR0 4YY
14/09/2026
02216256-0009